New Techniques for Future Accelerators III

High-Intensity Storage
Rings–Status and Prospects
for Superconducting Magnets

ETTORE MAJORANA
INTERNATIONAL SCIENCE SERIES
Series Editor:
Antonino Zichichi
European Physical Society
Geneva, Switzerland

(PHYSICAL SCIENCES)

Recent volumes in the series:

A Continuation Order Plan is available for this series. A continuation order will bring delivery
of each new volume immediately upon publication. Volumes are billed only upon actual ship-
ment. For further information please contact the publisher.

New Techniques for Future Accelerators III

High-Intensity Storage Rings—Status and Prospects for Superconducting Magnets

Edited by

Gabriele Torelli

University of Pisa
Pisa, Italy

Plenum Press • New York and London

Library of Congress Cataloging-in-Publication Data

Seminar on New Techniques for Future Accelerators (3rd : 1989 : Erice,
  Italy)
    New techniques for future accelerators III : high-intensity
  storage rings, status and prospects for superconducting magnets /
  edited by Gabriele Torelli.
        p.   cm. -- (Ettore Majorana international science series.
  Physical sciences ; v. 53.)
    "Proceedings of the tenth workshop of the INFN Eloisatron Project:
  Seminar on New Techniques for Future Accelerators III ... held
  October 16-24, 1989, in Erice, Sicily, Italy"--T.p. verso.
    Includes bibliographical references and index.
    ISBN-13: 978-1-4684-5861-9   e-ISBN-13: 978-1-4684-5859-6
    DOI: 10.1007/978-1-4684-5859-6
    1. Superconducting magnets--Congresses. 2. Storage rings-
  -Congresses. 3. Particle accelerators--Technique--Congresses.
  I. Torelli, Gabriele. II. Title. III. Series.
  QC761.3.S46   1989
  539.7'3--dc20   90-26665
                                                                 CIP

Proceedings of the Tenth Workshop of the INFN Eloisatron Project:
Seminar on New Techniques for Future Accelerators III: High-Intensity
Storage Rings—Status and Prospects for Superconducting Magnets,
held October 16–24, 1989, in Erice, Sicily, Italy

ISBN-13: 978-1-4684-5861-9

© 1990 Plenum Press, New York
Softcover reprint of the hardcover 1st edition 1990

A Division of Plenum Publishing Corporation
233 Spring Street, New York, N.Y. 10013

PREFACE

    A fundamental step towards gaining a deeper understanding
of our world is to increase the resolution of the investigative
instruments we use; i.e. to increase the energy, and hence to
decrease the wavelength, of the particles which constitute our
probes.

    Almost any substantial progress in our understanding of the
fundamental laws of Nature has been obtained when a new
generation of accelerators has allowed us to achieve a new
energy range. The new results have generated new questions,
thus encouraging us to construct new machines to reach even
higher energy levels.

    The relative energy gain from one generation of
accelerators to the next is progressively increasing. The
energy gain suggested by the theoretical predictions at the
time has usually been much greater than the value allowed by
our technical capabilities. But this smaller energy gain
permitted by accelerator technology improvement has generally
been sufficient up until now to bring about a substantial
increase in our knowledge.

    Hence a large increase in accelerator energy is very
important, and we know that this result can essentially be
obtained by developing some new device or some new approach.
Bearing this in mind we dedicated the Third Workshop of the
"INFN Eloisatron Project" to RF techniques looking at
Superconducting Cavities and Microwave Devices, and we now open
this Ninth Workshop – the third Seminar devoted to New
Techniques for Future Accelerators – to examine closely the
state of the art for the production of the other fundamental
component of any future accelerator : the Superconducting
Magnet.

    I am very glad to open this Workshop in this delightful
town of Erice and I hope that you, a very qualified group of
specialists, working together for a week in a location ideally
suited to the fruitful interchange of ideas between scientists,
will greatly improve our knowledge in this field and thus play
a part in its future development.

                                            Gabriele Torelli

CONTENTS

THE ELOISATRON

K. Johnsen

CH-1261 La Rippe

Switzerland

1. INTRODUCTION

The study of an exceptionally large proton-proton collider, the Eloisatron, has been initiated by Professor Zichichi. The goal is to reach 200 TeV centre-of-mass energy in a tunnel of 300 km circumference. The bending field of the collider will have to be about 10 T. The total bending length will in that case be 209.6 km/ring. The rest of the circumference will be used for focusing elements, correcting magnets, acceleration cavities, beam instrumentation, injection and extraction systems, and most important of all, the insertions for the interaction regions. It is tentatively proposed to house these in two major groups, each group occupying about 15 km. The facility will thus have the shape of a race track similar to the arrangement already proposed during the LSR study about 10 years ago and recently adopted in the SSC design. Each group could have three interaction points, as indicated in Fig.1.

The main rings will be fed from a cascade of synchrotrons (most likely three in succession) which in turn will be fed from a linear accelerator. This is also shown schematically in Fig.1. Here the last of the injector synchrotrons is assumed placed in the same tunnel as the main rings, and therefore does not appear as a separate ring, but other arrangements are also possible.

In the following are some comments on the main subsystems.

2. INSERTIONS

The main purpose of the experimental insertions is to transform the beam properties such that the crossing regions become very intense event sources while still maintaining accessibility for detectors and otherwise making the facility flexible and convenient to use.

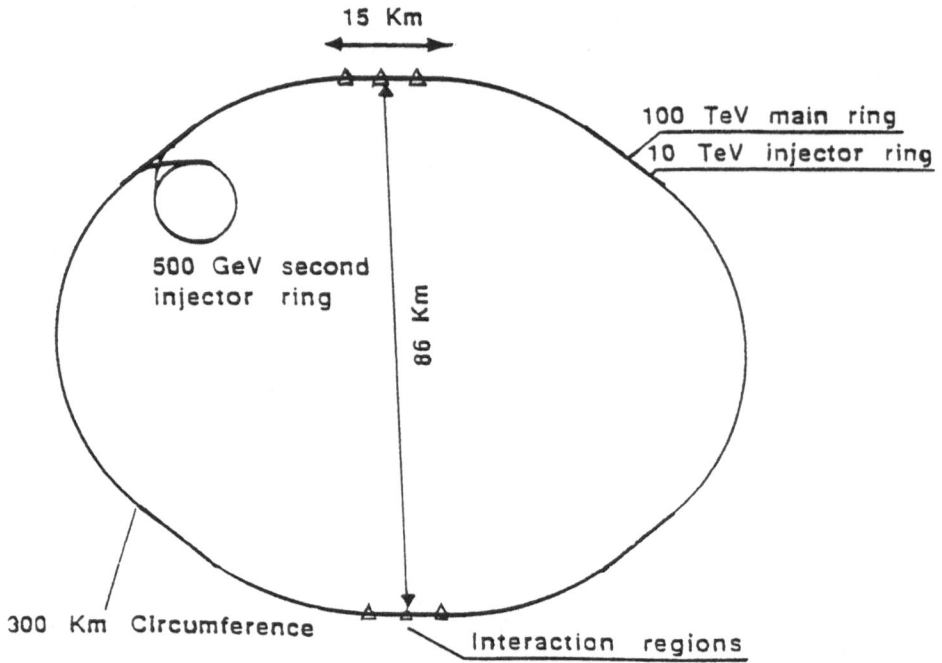

Fig.1. Possible ELOISATRON layout

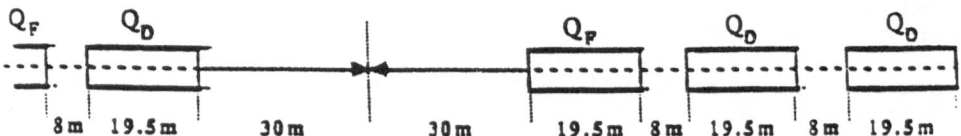

Fig.2. Anti-symmetric layout of experimental insertion
This arrangement would give:

$\beta_{max} = 18900$ m   for $\beta^* = 0.6$ m
$\beta_{max} = 4770$ m   for $\beta^* = 2.4$ m

These are often contradictory requirements, and compromises must be found, perhaps leading to different solutions for different interaction regions. The final arrangement will also depend much on detector developments over the next decade or so, for instance, on how beam-focusing systems can be integrated into detector components. It seems likely that interaction regions of very high luminosity will be wanted with the strongest possible focusing near the interaction points, for instance, a of about 1 m or less for luminosities above 10^{33} $cm^{-2}s^{-1}$.

In order to see whether this is feasible, simple scaling from the corresponding LHC configuration can be performed (CERN87-05), Only the triplets at either side of the interaction point will be considered since these are the most critical elements. Scaling gives a 75 m long triplet consisting of four 15 m long quadrupoles with 5 m between each. Their gradient becomes 350 T/m. The free space from the interacting point to the first quadrupole is 30 m and the system has an antisymmetric quadrupole arrangement about the interacting point. The β at the crossing point can be made 1.25 m and the maximum β in the triplet becomes 8250 m.

Recently Scandale has presented an improved insertion for LHC (LHC note 68). If this design is taken as a basis for scaling, different parameters and smaller $\beta*$ are also obtained for an ELOISATRON insertion. The result is a triplet length of 102 m. Each triplet is made of four 19.5 m long quadrupoles, again with 350 T/m gradient. The free space would still be ±30 m. This structure is shown in Fig.2.

Thus a β of less than 1 m seems realistic, at the expense of very large β_{max}. As will be seen later, this means that the machine can be tuned to the lowest β (and therefore highest luminosities) only in the high-energy range. Near injection energies one will have to be content with more moderate values of $\beta*$. In all this it is assumed that chromaticity corrections are possible to make the dynamic aperture at least as large as the physical aperture. More studies are needed both on the detector side and the machine side to arrive at overall optimum insertions. A particular problem will be to incorporate a system for dumping the 100 TeV protons with stored energies in the gigajoule range.

3. MAIN RINGS

3.1 Lattice

The main rings should provide two colliding proton beams of 100 TeV each with the highest feasible luminosity in the interaction regions. A total circumference of 300 km will require a bending field of about 10 T.

Simple scaling from existing designs of similar smaller projects gives a good starting point for further analysis. An example is given in Table 1.

TABLE 1 - LATTICE PARAMETERS (1 RING)

Length of normal period	200 m
Phase advance per period	$\pi/3$
Betatron wavelength	1200 m
Maximum betafunction in the normal lattice	340 m
Bending angle per normal period	4.7 mrad
Number of quads per period	2
Effective length of each quad	13.6 m
Number of quadrupoles (without insertions)	2664
Maximum dispersion in the normal lattice	1.5 m
Maximum dipole field	10 T
Bending radius	33356 m
Number of dipoles per normal period	12
Effective dipole length	13.1 m
No of dipoles	15984

3.2 Performance evaluation and performance-related parameters

Once the energy has been fixed, the next most important parameter is the luminosity given by

$$L = N_p^2 f_b / (4\pi\sigma^2) = N_p^2 f_b v / (\beta^* \varepsilon)$$

Here N_p is the number of particles per bunch, f_b is the average bunch frequency, s is the r.m.s. beam radius at the crossing points, and ε is the normalized emittance defined by

$$\varepsilon = 4\pi\sigma v^2 / \beta^*$$

β^* is the beta value at the crossing point.

These equations illustrate where to put emphasis to obtain high luminosity. There are, however, a few limitations; the most important ones for the design can be listed as follows:

i) The detectors have difficulty in discriminating between events if the bunches cross more frequently than say, once every t_b = 25 nsec. t_b is taken as a limit on the bunch frequency

$$f_b < t_b^{-1} = 40 \text{ MHz}$$

or 7,5 m between bunches.

ii) The beam-beam tune shift parameter ξ must be kept smaller than a given number. ξ is given by

$$\xi = N_p r_p / \varepsilon$$

where r_p is the classical proton radius. For pp interactions the total tune shift over all interaction regions should be less than 0.01, which gives

$$\epsilon < 0.01/6 = 0.00167$$

for each of six interaction regions.

iii) It is difficult to make $\beta*$ small. For SSC at 20 TeV it is assumed that $\beta*$ can be made 0.5 m. For 100 TeV we assume $\beta* \approx 1m$, i.e., approximately in the middle of the range arrived at in section 2.

iv) Synchrotron radiation at 100 TeV becomes a serious load on the cryogenic system of the magnets. More than, say, 2 W/m on the vacuum pipe is undesirable, and one should trap much of this on separately cooled surfaces.

v) The normalized emittance from present-day proton accelerators is $5\pi \times 10^{-6}$ m or larger. However, since iv) above will be imposing a limit on the circulating current far below the capability of an accelerator, it is assumed that the emittance can be correspondingly reduced to say

$$\epsilon \approx 0,75 \; \pi \times 10^{-6} \text{ m}$$

Some of the conditions may change as the various technical solutions are being studied. Examples, however, have been worked out and a few iterations have resulted in tentative numbers presented in Table 2.

TABLE 2 Some tentative performance parameters for Eloisatron

Energy per beam	100 TeV
Number of bunches (per beam)	39600
β-value at interaction point	1.25 – 0.6 m
Normalized emittance	$0.75\pi \times 10^{-6}$ m
r.m.s. beam radius at inter. point	$1.25 - 0.9 \times 10^{-6}$ m
Circulating current	16.43 mA
Particles per bunch	2.56×10^9
Beam-beam tune shift per crossing (with 6 active crossings)	1.67×10^{-3}
Bunch spacing	$25 \times 10{-9}$ s
Stored beam energy	1.623×10^9 J
Luminosity	$0.9 - 1.8 \times 10^{33}\text{cm}^{-2}\text{s}^{-1}$
Energy loss per turn due to synchrotron radiation	23.24 MeV
Radiated power (per beam)	385 kW
Power per unit bending length of one beam	1.84 W/m
Transverse em. damping time	1.2 h

3.3 Beam-Pipe Aperture

The physical aperture of an accelerator constitutes a hard limit on the beam which cannot be changed once the machine has been built. For this reason the aperture requirement should be estimated on the basis of conservative assumptions.

The choice of injection energy will probably be in the 5-10 TeV range. In the following aperture estimates the lower figure of 5 TeV is assumed.

In section 3.2 a normalized emittance of $0.75\pi \times 10^{-6}$ m was assumed for performance estimates. This may, however, be optimistic, at least in the early operational phase. For aperture determinations we therefore take the more common figure of $5\pi \times 10^{-6}$ m.

The beam does not need more room than 4σ at injection. This would give good beam conditions of 8σ above 20 TeV, the lowest energy of interest for physics.

The closed-orbit deviations will depend on the quality of the magnets, the beam observation and the correction system. Lacking detailed information LEP values are scaled. The result is shown in Table 3.

TABLE 3 - APERTURE EVALUATION

Max. beam radius (4σ) at injection	2.6 mm
Corrected closed-orbit error	± 4 mm
Radial extent of bunch	± 0.4 mm
Sagitta	± 0.4 mm
Needed vertical aperture	± 6.6 mm
Needed horizontal aperture	± 7.4 mm
Aperture of vacuum chamber	± 8.0 mm

In the last figure an allowance is made for coating and tolerance of ± 0.6 mm has been added.

We assume that the low-β insertion will only be fully activated when an emittance of $0.75\pi \times 10^{-6}$ m has been achieved. This will then allow $8\sigma = 2.5$ mm in the maximum β point at 20 TeV and $\beta*$ 1 m. Above 40 TeV the $\beta*$ can be lowered to 0.6 m with 8σ 2.5 mm at the maximum β point.

3.4 Coil aperture

Additional space is needed for vacuum chamber wall, correction windings, thermal insulation and cooling pipes to remove the heat from synchrotron radiation.

The last problem is more pronounced at the Eloisatron than other projects under study.

Estimates can be made and Fig.3 represents an example.

3.5 COMMENTS ON SOME OF THE TECHNICAL COMPONENTS

3.5.1 Magnets

Some remarks can be made about various possibilities:

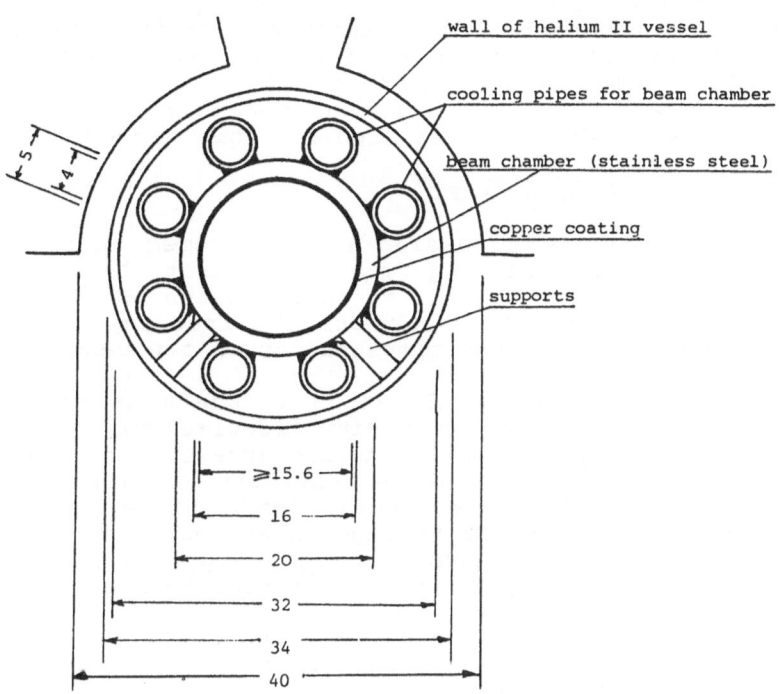

Fig.3 Example of coil aperture

1) Niobium – titanium is the available winding material: By going to 1.8 °K one can reach 8-10 T. The optimum cost vs field will have to be studied.

II) If NB_3Sn is chosen, 10^{-12} T may be reached at normal liquid helium temperature. Much development on metallurgical processes for wire production and on the thermal treatment of the insulated windings is needed to arrive at a satisfactory design. Such work goes on at many places in the world, Europe included.

iii) With the recent dramatic development of "warm" superconducting materials, there is hope that such materials will become available for accelerator magnets in the future. This must be followed closely. However, such solutions will introduce new problems. For instance, the advantage of cryogenic pumping is lost and distributed pumping of very small transverse dimensions will have to be developed.

iv) There are two possibilities for the magnetic circuits of the two rings. The conventional approach is to make the two rings magnetically independent, i.e. each magnet has its own return yoke. Another approach is to let each magnet gap take the return flux from its neighbour. This "two-in-one" approach saves some space, superconducting material, steel and cryostat. But since a serious restriction is imposed on flexibility (difficulty to tune and operate the rings separately) the SSC Design Group has chosen independent rings. The same is recommended for the ELOISATRON since the savings mentioned above are small. However, it may still be desirable to put both magnets in the same cryostat (introducing only a weak operational dependence).

The Eloisatron magnet development can profit from the on-going R&D on the SSC and more particularly from the R&D needed for the high-field magnets of the LHC, however, since the Eloisatron magnet will have its own special problems a dedicated R&D activity should be created for this project.

3.5.2 - <u>Radiofrequency Systems</u>

For this preliminary design it is assumed that the beam be injected into the main ring from an injector chain that will produce beam properties similar to those of the LHC. This leads naturally to a choice of 400 MHz for the frequency of the accelerating systems (i.e. ten times the bunch frequency).

The time taken for acceleration in the main ring should be short compared with the luminosity lifetime during an experimental run (flat top), Further, during acceleration the bucket size should be considerably larger than the longitudinal emittance of a bunch. These two requirements favour a large RF voltage. However, a large RF voltage is expensive and a reasonable compromise must be sought. At top energy the protons lose about 25 MeV per turn due to the synchrotron radiation.

For acceleration the energy gain ought to be a few times larger, say 7 5 MeV/turn during most of the ramping time . This leads to about 20 min ramping time.

For a first approximation, an RF peak voltage of twice this seems a reasonable starting point for an analysis. Consequently an RF peak voltage of u = 150 MeV is taken for the following calculations.

A constant ramping rate cannot be maintained all the time since the bucket size would become uncomfortably small at low energy. Thus the ramping rate is tailored such that the bucket size stays constant and equal to 25 eVs during the entire ramping. This leads to an energy gain as shown in Fig.4.

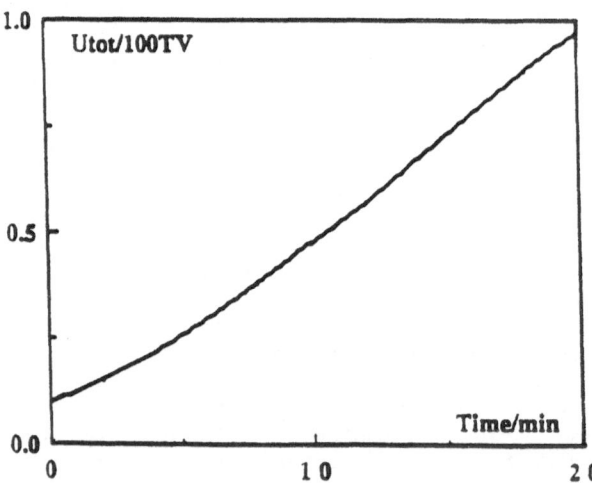

Fig. 4. Total energy as function of time during ramping

The rate is lowered at low energy and also somewhat near the top energy, whereas it is increased above 75 MeV/turn around the middle. The total time from 10 TeV to 100 TeV is still only about 20 min.

The RF requirements can be compared with LEP at 55 GeV. The peak voltage per ring for Eloisatron is 45% of that of LEP 55. This means that the installed RF power per ring will have to be at least 7.2 MW, which is large enough that it is worth considering superconducting cavities, and their development at CERN and at DESY must be followed closely .

With the above information Table 4 is derived for the main parameters of the RF system.

TABLE 4 - MAIN RF PARAMETERS (ONE RING)

Frequency (MHz)	4⁰⁰
No. cavities	56
No. klystrons	8
Length of active RF structure(m)	120
Installed power (MW)	8
Peak voltage per turn (MW)	150
Synchrotron radiation at 100 TeV (MeV)	23.3

4. CONCLUDING REMARKS

This preliminary study demonstrates that an Eloisatron, with performance as given in this paper, seems feasible by extrapolation and "stretching" of present-day technologies. The technological development is fast-moving for some of the crucial accelerator components like magnets and radio-frequency systems, where in particular the latest development of superconductivity may offer new prospects. The Eloisatron design should take full advantage of such development and incorporate new ideas as they become realistic.

5. ACKNOWLEDGEMENT

Dr Werner Hardt has made important contributions to the analysis of the RF system. He has also helped in the general editing of the paper. Other colleagues have also been willing partners in discussions of problems related to this report, and the author wants to express gratitude for all their assistance.

THE BASIS FOR THE DESIGN OF SUPERCONDUCTING ACCELERATOR MAGNETS

Peter Schmüser

II. Institut für Experimentalphysik, Universität

Hamburg

SUMMARY

The basic design principles of superconducting accelerator magnets were covered in a series of three talks following closely the lectures by K.H. Meß and myself at the Course on Superconductivity in Particle Accelerator[1], Hamburg 1988. Since a detailed report is contained in the proceedings of the Course, I want to restrict myself here to a brief summary of the topics except for those points where important new material has been added.*

---

1. K.-H. Meß and P. Schmüser, Proceedings of the Course on Superconductivity in Particle Accelerators, Hamburg 1988, CERN yellow report 89-04 (1989) and DESY report HERA 89-01 (1989)

* New results have been presented on mechanical accuracies, forces and quench studies.

MECHANICAL ACCURACIES

The quality control measurements which are performed on all HERA dipole and quadrupole magnets yield a wealth of information on the accuracies which can be achieved in a large-scale industrial production of superconducting magnets. One of the most stringent tests on the precision of the coil cross section and of the placement of the current conductors is provided by the multipole measurements. Figure 1 shows the normal and skew multipole coefficients of the HERA dipoles at the nominal current of 5000 A. Plotted are the average values of 200 magnets with their rms standard deviations. Most of the coefficients are very small and well within the limits of $\pm 0.5 \cdot 10^{-4}$, which are used in the particle tracking program to determine the dynamic aperture of the HERA storage ring. Two coefficients show a larger scattering: the normal sextupole b_3, which is particularly sensitive to slight changes in the key angles of the coils sections, and the skew quadrupole a_2, which may arise from an up-down asymmetry of the coil. The dipoles will be sorted in the ring to minimize the sextupole effects.

The general conclusion from the data in Fig.1 is that a very high mechanical precision has been achieved in both the Italian dipoles (made by ANSALDO and ZANON) and the German ones (made by ABB). In the companies the collared coils were measured at room temperature before their installation in the iron yoke and cryostat. If the sextupole turned out too large, the collars were opened, shims were added and then the collars were closed again. In order to minimize top-bottom asymmetries the two half coils in a dipole were matched with respect to their elastic moduli.

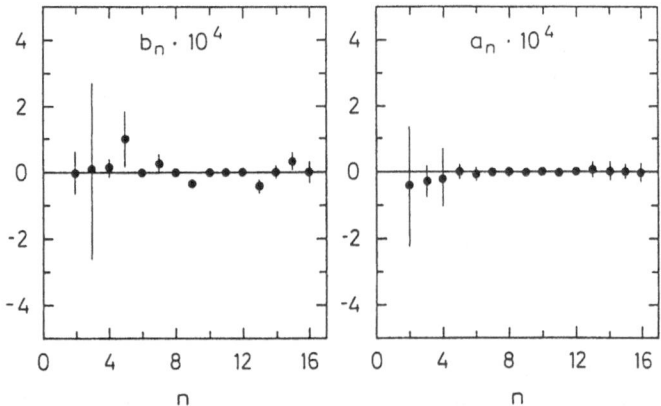

Fig. 1.   The normal (b_n) and skew (a_n) multipole coefficients of the HERA dipoles at 5000 A, corresponding to a proton energy of about 800 GeV. Plotted are the average values of 200 magnets with rms standard deviations. The data have been averaged over the whole length of the magnets, including the coil heads.

From measurements made at ANSALDO on the stress-strain relation in cured half coils one can derive an elastic modulus $E = (27 \pm 4)$ GPa.

In Fig.2 the normal and skew multipoles from all quadrupoles are plotted. The higher-order fields relate to the quadrupole field at our reference radius $r_0 = 25$ mm. The tolerable limits are $\sigma = \pm 5 \cdot 10^{-4}$ (the quadrupoles are shorter and weaker than the dipoles). All coefficients are well within the tolerable range. There are a few nonvanishing poles from coil geometry: b_6, b_{10} and in particular b_{14}. The agreement between the predicted and measured values is remarkable and illustrates both the quality of the coil manufacturing and of the measuring devices.

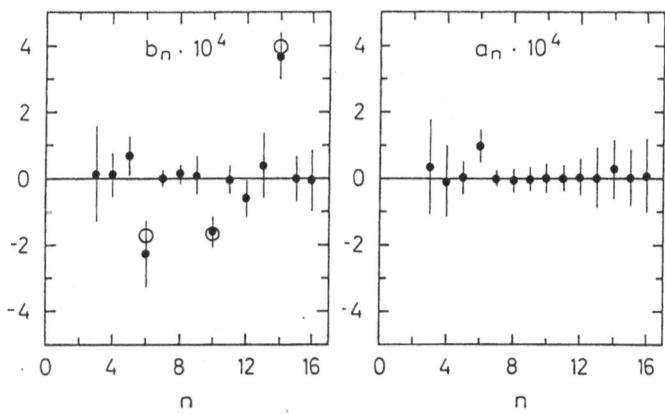

Fig. 2.   The normal and skew multipole coefficients of 225 quadrupoles at 5000 A. The theoretical values of the non-vanishing poles b_6, b_{10}, b_{14} are indicated by open circles.

In the following I summarize further measurements on the geometrical accuracy of the dipoles.

In the magnetic length a small but systematic difference is observed between the ABB and the ANSALDO-ZANON dipoles

$$l_{mag} = 8.8335 \pm 0.0020 \text{ m (ABB)}$$
$$8.8257 \pm 0.0017 \text{ m (ANSALDO-ZANON).}$$

This is due to small differences in the tooling and the wire tension during coil winding.

The magnetic field normalized to the current is a check on the radial coil dimensions. The numbers are

$$B/I = 0.9336 \pm 0.0005 \text{ T/kA} \quad \text{(ABB)}$$
$$0.9328 \pm 0.0005 \text{ T/kA} \quad \text{(ANSALDO-ZANON).}$$

The angle φ between the magnetic field and the direction of gravity is determined with two orthogonal Hall probes connected to a gravitational sensor. This is a check on the alignment of the coil inside the cryostat. Figure 3 shows an example of such a measurement. Averaged over the length of the magnet and over all magnets from each manufacturer the following numbers are obtained

$$\phi = 1.57 \pm 1.75 \text{ mrad} \quad \text{(ABB)}$$
$$0.07 \pm 1.20 \text{ mrad} \quad \text{(ANSALDO-ZANON)}.$$

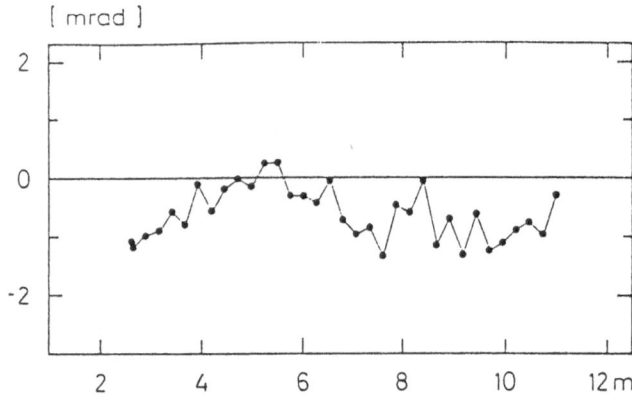

Fig. 3.   Field direction with respect to gravity along the axis of a dipole. The data points are reproducible to within 0.1 mrad, so the observed structure is a property of the magnet.

The same sensor is used to determine the twist of the magnet. The twist per m is

0.02 ± 0.31 mrad/m   (ABB)
0.09 ± 0.24 mrad/m   (ANSALDO-ZANON).

These numbers are well within the tolerable limits.

The HERA quadrupole coils with collars and yoke are contained in a strong stainless steel tube. Precisely mounted bars provide for the accurate centering and angular alignment of the coil (Fig. 4). In the companies (ALSTHOM in France, NOELL in Germany) the cold part is aligned inside the vacuum vessel by mechanical means. In the cryogenic measurements at DESY the magnetic axis of the quadrupole and the field direction is determined with a "stretched-wire" system. A thin wire is stretched through the magnet parallel to the axis. Together with an external wire it forms an induction loop. The wire is precisely moved horizontally or vertically and the induced voltage is integrated. From a series of such measurements the quadrupole axis, field direction and the integral of the gradient ∫gdl can be computed. The measured deviations Δx (horizontal) and Δy (vertical) between the magnetic axis and the prealigned axis are

$$\Delta x = - 0.02 \pm 0.36 \text{ mm}$$
$$\Delta y = - 0.38 \pm 0.33 \text{ mm}$$

The nonvanishing average value of Δy implies that the vertical position of the cold mass inside the vacuum vessel changes by 0.4 mm during cooldown. The magnetic measurements are used to align the quadrupoles in the HERA ring.

The angular accuracy of the prealignment is good. The average angle between the field direction and gravity is

$$\phi = 1.50 \pm 1.18 \text{ mrad.}$$

Again the magnetic data are used for the installation in the accelerator.

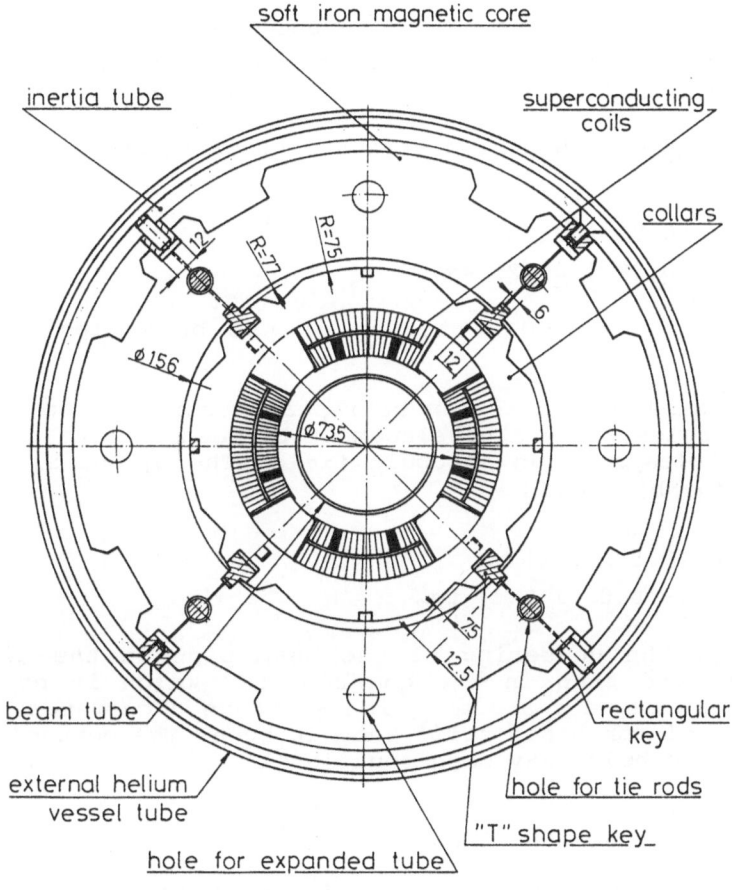

Fig. 4.   Cross section of the HERA quadrupole coil with stainless steel collars and soft iron yoke. The mechanical precision is provided by a strong support tube ("inertia tube").

A very good accuracy has been achieved within the cold part. The stainless steel support tube houses also the correction dipole. The measured angle between the correction dipole field and the quadrupole field is $- 0.17 \pm 0.62$ mrad.

A variety of correction coils are mounted on the cold beam pipes inside the main magnets: quadrupole, sextupole and ten-pole coils inside the dipoles, twelve-pole coils inside the quadrupoles. A good centering of the beam pipe is important. We have developed a method to determine this by magnetic measurements. Consider the quadrupole magnet. At low excitation the persistent currents generate a relatively large twelve-pole field (typically $b_6 = - 25 \cdot 10^{-4}$ at 250 A). The current in the twelve-pole coil is adjusted now so as to compensate this field. If the correction coil is not exactly centered inside the quadrupole, the twelve-pole field will be compensated but there will be a feed-down to a ten-pole field

$$b_5 = 5 \cdot \frac{\Delta x}{r_0} \, b_6 \qquad\qquad a_5 = 5 \cdot \frac{\Delta y}{r_0} \, b_6 \; .$$

Here, b_6 is the value of the uncompensated twelve-pole coefficient, Δx, Δy are the horizontal and vertical displacements of the beam pipe coil and $r_0 = 25$mm the reference radius.

Figure 5a shows the distributions of b_5 and a_5. From these we compute

$$\Delta x = (0.01 \pm 0.20) \text{ mm}$$
$$\Delta y = (0.08 \pm 0.22) \text{ mm}$$

indicating very good centering of the beam pipe inside the quadrupoles.

The same procedure can be repeated for the ten-pole coils in the dipoles. Here a displacement between correction coil and main coil leads to an octupole field. The b_4 and a_4 distributions (Fig.5b) yield

$$\Delta x = (0.5 \pm 0.6) \text{ mm}$$
$$\Delta y = (- 0.03 \pm 1.1) \text{ mm}.$$

Obviously, the centering of the beam pipe in the dipoles is less accurate than in the quadrupoles, which is no surprise since these magnets are much longer, have a curvature and lack the precisely machined stiff support tube. The octupole fields, however, are below any dangerous level.

FORCES ON COLLARS AND YOKE

The HERA dipole coils are confined by aluminium collars which have a circular outer shape in the unloaded case. After collaring the coil is loaded with a high prestress which amounts to 65 MPa at 4 K. The collars are correspondingly deflected to the shape of an upright ellipse (Fig. 6a). If the

16

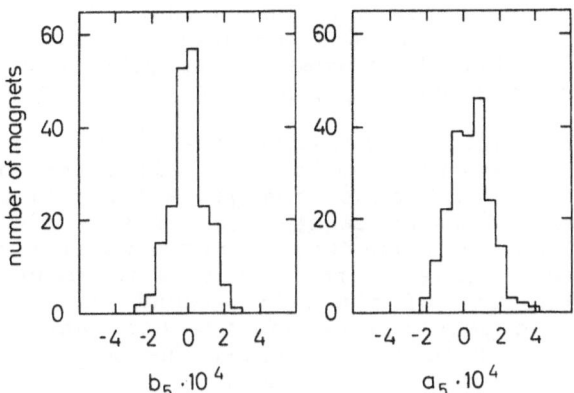

Fig. 5a. Normal (b_5) and skew (a_5) decapole coefficients measured in the HERA quadrupoles at I = 250 A. The current in the 12-pole coil was adjusted to compensate the persistent current 12-pole field of the quadrupole magnet.

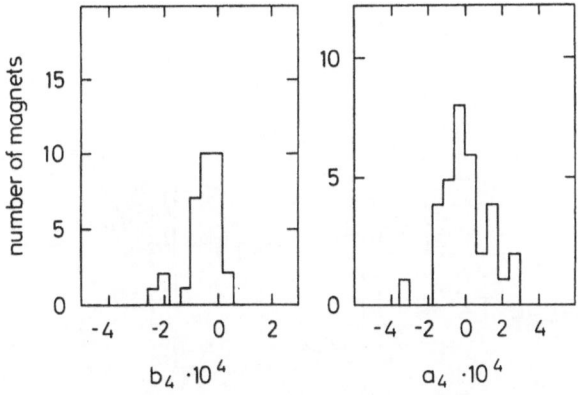

Fig. 5b. Normal (b_4) and skew (a_4) octupole coefficients in the HERA dipoles at I = 250 A. The current in the 10-pole correction coil was set to compensate the persistent current 10-pole field of the dipole.

coil is excited to its nominal field of 4.68 T, the Lorentz forces deform the collared coil to an ellipse with the large half axis in the horizontal direction (Fig. 6b).

The collars are interlocked with the yoke by four notches which fit into corresponding grooves in the yoke laminations. There is a clearance between collars and yoke which is adjusted such that for fields below 6 T the Lorentz forces are taken up by the collars alone. Above 6 T the deflected collars touch the iron yoke and transfer the forces to the yoke so no further deflection will occur since the yoke is much stiffer.

In the HERA dipoles the iron yoke is assembled from two half yokes with a vertical split. The half yokes are connected by two weld joints which close the gap at room temperature and also at 4 K. Two 3.5 mm thick stainless steel half shells surround the yoke and are connected by two simultaneously applied weld joints. The shrinkage after welding is about 0.8 mm thus providing a tight fit of the tube around the yoke. The weld joints between the yoke halves and the half shells are strong enough that either of them can support the Lorentz forces up to the highest excitation of the magnet.

Particular attention has been paid to the interface between the stamped aluminium laminations of the collars and the soft iron laminations of the yoke. A U-shape bronze sheet of 0.5 mm thickness is inserted in between to provide a well-defined sliding plane. In some SSC prototype magnets the notches of the stainless steel collars were in direct contact with the grooves leading to slip-stick motion and sudden jumps caused by the longitudinal forces.

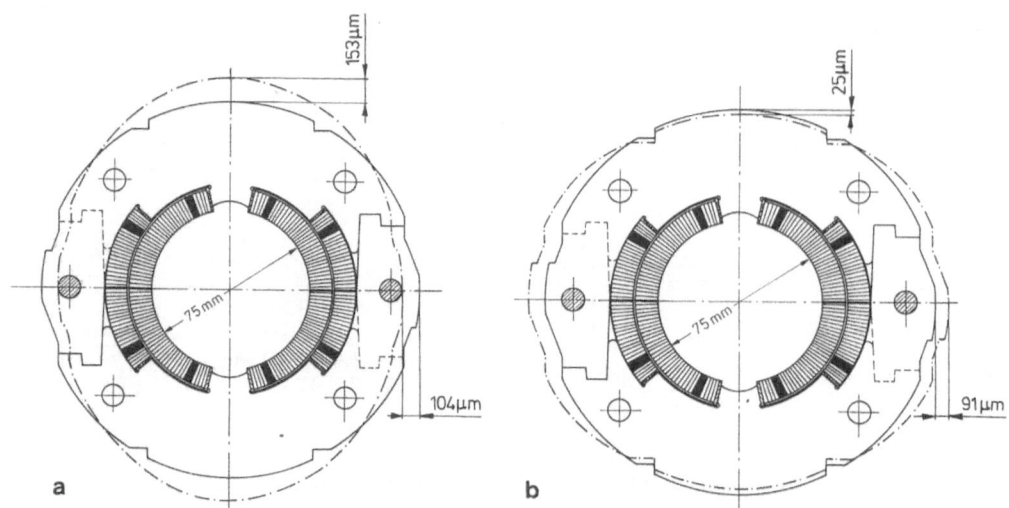

Fig. 6.   Calculated deformation of the collared HERA dipole coil at 4 K: a) without magnetic field, b) at 6 T.   The collar material is AlMg4.5Mn(G35) with σ_{02} = 270 MPa and a yield strength of 350 MPa. The maximum calculated stress in the collar is 150 MPa. The calculation was performed by G. Meyer of DESY. In a test setup with a mechanical load applied to a short stack of collar laminations the measured and computed collar deflections agreed to within 5%.

The longitudinal Lorentz forces acting on the coil ends amount to about 150 kN in the HERA magnets. We have followed the philosophy that these forces should be supported where they arise. The coil ends are therefore confined in the longitudinal direction by stainless steel end plates which rest with zero clearance against the yoke end plates. In principle the super-conducting cable is strong enough to take up the longitudinal forces by itself but then the coil would lengthen by about 3 mm in the HERA case. The motion might trigger quenches in particular if there were no well-defined sliding medium between collars and yoke.

QUENCH STUDIES

The SSC magnet group has performed systematic quench studies[2] on 17 m long dipoles which were equipped with a lot of diagnostics, in particular spot heaters and voltage taps. Very interesting and useful data on quench propagation and heating of the coil have been obtained.

Figure 7 shows the longitudinal quench propagation velocity plotted against the ratio of operating to critical current density. Surprisingly large velocities of typically 100 m/s are observed. The velocities are higher than expected from a simple adiabatic model. Probably a preheating by rapidly propagating warmer helium plays a role.

The transverse (turn-to-turn) propagation velocity (Fig. 8) is of course much smaller due to the insulation in between the windings but is still typically 100 turns per second or more.

By applying a voltage pulse to a spot heater a short section of a winding was quenched. The temperature of the winding was derived from the resistive voltage developing between the voltage taps at this section. The results are summarized in Figs. 9 and 10. For an induced quench in the inner coil layer at the nominal current of 6500 A a hot spot temperature of 150 K was observed, in good agreement with a calculation. Higher temperatures were found for quenches in the outer coil layer (Fig.10). Here the quench propagation is slower. There is an important relation between the time-integral over the squared current, $\int I^2 dt$, and the maximum temperature in the coil (see e.g. Ref.1). Figure 11 shows the hot spot temperature as a function of the number of "MIIT's" (10^6 A^2 s). Experimental data and theoretical predictions are in good agreement.

In a string test of three HERA dipoles and two quadrupoles we have investigated whether quenches can be triggered by eddy current heating. By choosing appropriate dump resistors the current was decreased with a well-defined time constant. In several cases with high initial current and short decay times quenches were in fact observed, probably caused by the eddy current heating of the copper in the cable and of the aluminium collars.

A mathematical model was developed[3] to simulate the experiments. A simple adiabatic calculation turned out too pessimistic. The heat capacity of the helium enclosed in the cable had to be taken into account but also the heat transfer through

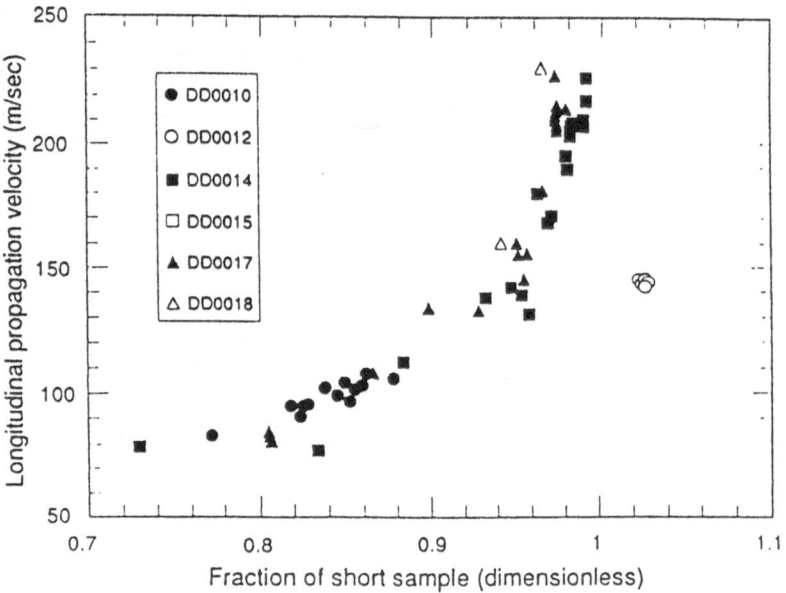

Fig. 7. Longitudinal quench propagation velocity in SSC dipoles (Ref. 2).

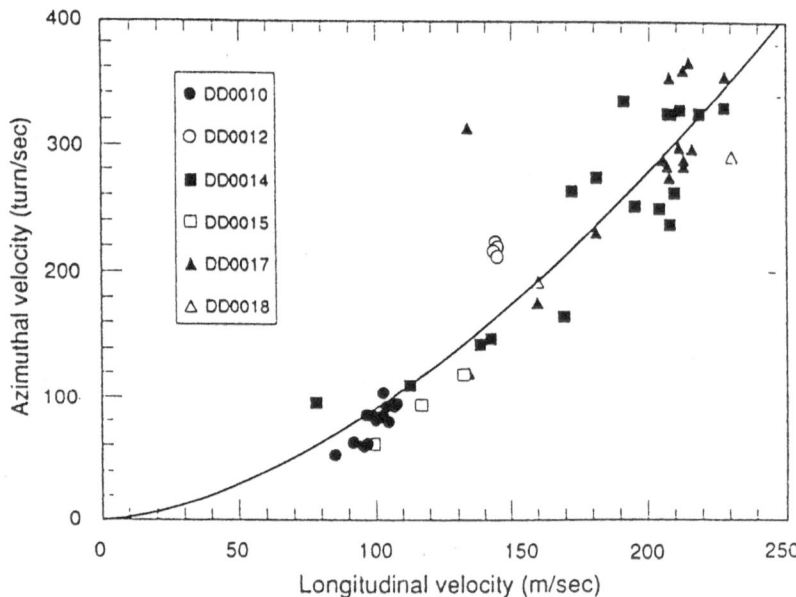

Fig. 8. Transverse (turn-to-turn) propagation velocity as a function of the longitudinal velocity (Ref. 2).

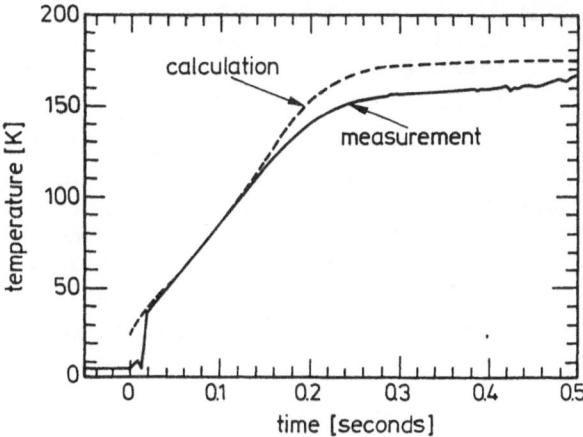

Fig. 9.   Evolution of the hot-spot temperature during a quench induced
           at 6500 A on turn 1 of the inner coil.

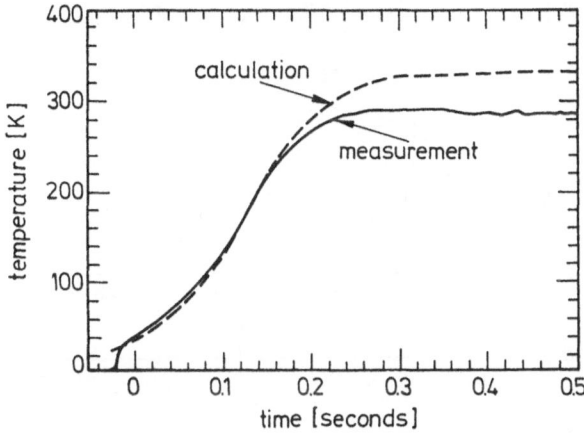

Fig. 10.   Evolution of the hot-spot temperature during a quench induced
           at 6500 A on the outer coil turn 20.

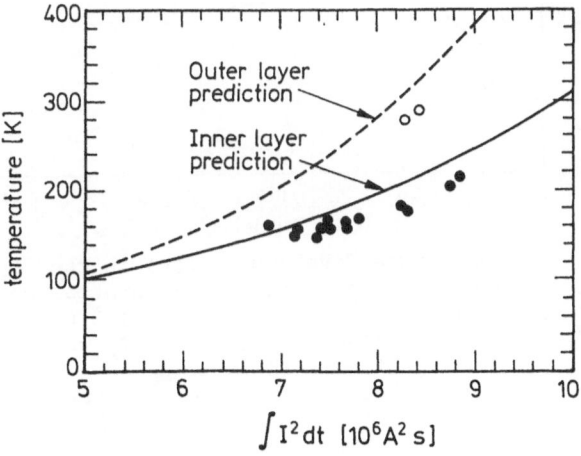

Fig. 11.   Hot-spot temperature as a function of the $\int I^2 dt$, measured in
           $10^6 A^2 s$ (so-called "MIIT")

the cable insulation to the helium between coil and beam pipe. Figure 12 shows the model prediction on the temperature development in the coil caused by an exponential current decay $I(t) = I_0 \exp(-t/\tau)$ with an initial current $I_0 = 4500$ A. For a decay time $\tau = 6$ s a quench was actually observed after 2.2 s whereas for $\tau = 11.3$ s no quench was found. By adjusting the unknown heat transfer coefficient from the coil through the insulation to the helium it was possible to reproduce these and other observations quite well. Figure 12a shows that for $\tau = 6$ s the coil temperature in fact exceeds the critical temperature after 2.4 s, whereas for $\tau = 11.3$ s (Fig.12b) it stays always below T_c (j,B).

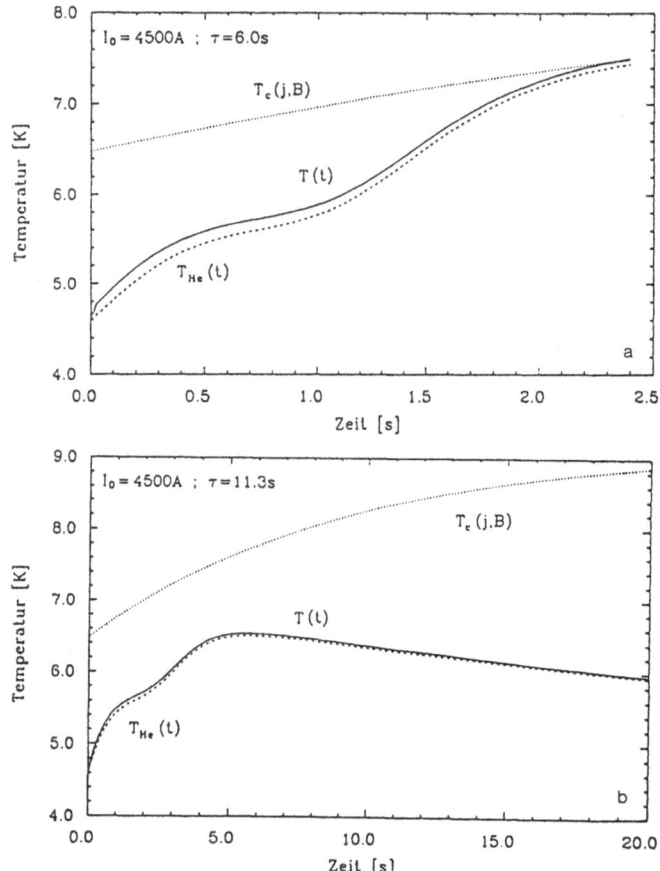

Fig. 12.   Simulated temperature development $T(t)$ in the dipole coil caused by eddy current heating during an exponential current decay $I = I_o \exp(-t/\tau)$: a) $\tau = 6.0$ s, b) $\tau = 11.3$ s. Plotted as a dotted curve is also the critical temperature T_c which depends implicitly on time as the magnetic field is dropping. When the coil temperature T exceeds T_c a quench is to be expected.

In the same setup the frequency response and transmission line behaviour of a chain of superconducting magnets was studied. The magnets have an inductivity of 60 mH, almost vanishing resistance (some effective resistance is present through inductive coupling with eddy current loops), and a sizable capacity (about 50 nF) against ground. So they should act like a resonant circuit and in fact a pronounced resonance with phase transition from +90° to -90° is observed in the string (Fig.13a). Fortunately, the resonance can be damped easily. The resistors which are needed for our quench detection system (see Ref.1) damp the resonance completely (Fig.13b).

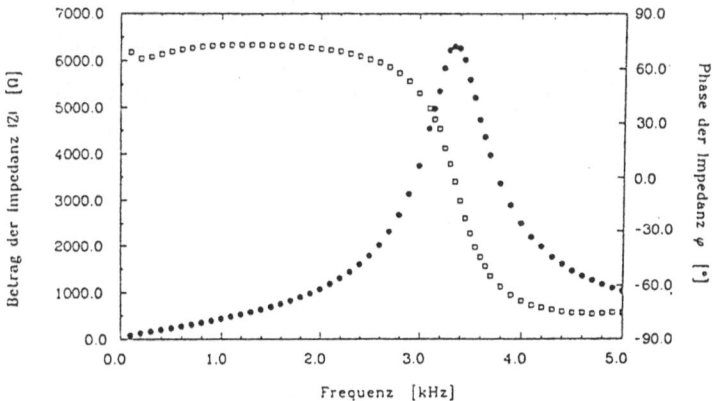

Fig. 13a. The complex impedance (magnitude and phase) of a system of superconducting magnets, measured as a function of the frequency

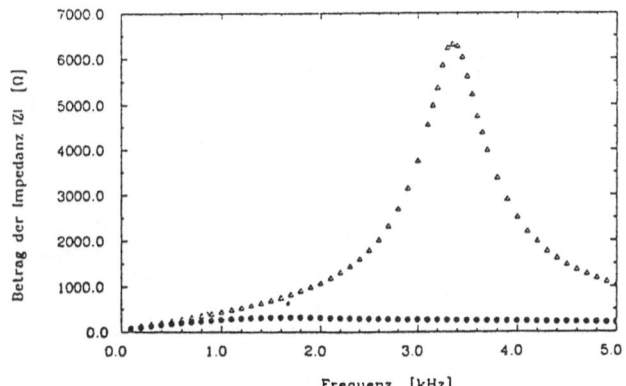

Fig. 13b. Damping resistors between the potential taps of the dipole half coils damp the resonance completely (black dots).

REFERENCES

1.   K.-H. Meß and P. Schmüser, Proceedings of the Course on
     Superconductivity in Particle Accelerators, Hamburg 1988,
     CERN yellow report 89-04 (1989) and DESY report HERA 89-
     01 (1989)

2.   A. Devred, M. Chapman, J. Cortella, A. Desportes, J.
     DiMarco, J. Kaugerts, R. Schermer, J.C. Tompkins, J.
     Turner, J.G. Cottingham, P. Dahl, G. Ganetis, M. Garber,
     A. Gosh, C. Goodzeit, A. Greene, J. Herrera, S. Kahn, E.
     Kelly, G. Morgan, A. Prodell, E.P. Rohrer, W. Sampson, R.
     Shutt, P. Thompson, P. Wanderer, E. Willen, M.
     Bleadon,B.C. Brown, R. Hanft, M. Kuchnir, M. Lamm, P.
     Mantsch, P.O. Mazur, D. Orris, J. Peoples, J. Strait, G.
     Tool, S. Caspi, W. Gilbert, C. Peters, J. Rechen, J.
     Royer, R. Scanlan, C. Taylor, J. Zbasnik,
     SSC report SSC-234 (1989) and Contribution to the 11th
     International Conference on Magnet Technology, Tsukuba,
     Japan, 1989

3.   B. Blau, Diploma thesis, University of Hamburg, DESY
     report HERA 89-19 (1989)

PERSISTENT CURRENT EFFECTS IN SUPERCONDUCTING ACCELERATOR

MAGNETS

Peter Schmüser

II. Institut für Experimentalphysik, Universität

Hamburg

ABSTRACT

Measurements and calculations are presented on the sextupole and decapole components in dipole magnets and the 12-pole and 20-pole components in quadrupoles. The data show a strong current dependence and a characteristic hysteresis behaviour. Good agreement is found with model calculations which are based on eddy currents in the niobium-titanium filaments of the superconducting cable. The influence of these currents on the dipole or quadrupole field has also been studied and is well reproduced by the model. The persistent current effects show an almost logarithmic time dependence which is possibly caused by flux creep in the superconductor. The creep rates vary considerably from magnet to magnet and are quite different for superconductors from different manufacturers.

INTRODUCTION

In my talk I want to report the data on persistent current multipoles[1] and their time dependence[2] obtained from measurements on a large number of superconducting dipole and quadrupole magnets for the proton-electron collider HERA[3] at Hamburg. The proton storage ring of HERA has a top energy of 820 GeV. The required dipole fields of 4.68 T and quadrupole gradients of 91.2 T/m are achieved with a current of 5027 A in the coils. During the injection of the proton beam at an energy of 40 GeV, the superconducting HERA magnets are excited to only 5% of their nominal field. The field distortions from persistent currents are appreciable and require the use of correction coils[4], in particular for the sextupole which has a strong influence on the chromaticity of the accelerator. But also the 10-pole field in the dipoles and the 12-pole field in the quadrupoles have to be compensated to ensure a sufficiently large dynamic aperture of the accelerator[5].

A precise knowledge of the persistent current fields is indispensable to enable appropriate correction schemes during the injection and initial acceleration of the proton beam. For this reason, detailed measurements are performed on each magnet before its installation in the ring. A theoretical understanding is needed as well because the field distortions depend on several parameters like the helium temperature and the previous current cycles which may be different in the test facility and in the accelerator. Following previous work by M.A. Green[6] and J.-L. Duchateau[7] a computer code has been developed at DESY which reproduces the data with remarkable success[8].

A short description of the HERA magnets and the multipole expansion is given in Section 2. The magnetization current model is described in Section 3 and the model predictions are compared to the experimental data. Section 4 is devoted to the time dependence measurements and their implications for the storage ring operation. In Section 5 I want to make some remarks on the behaviour of superconductors in magnetic fields which lead to a qualitative understanding of the observed time dependence.

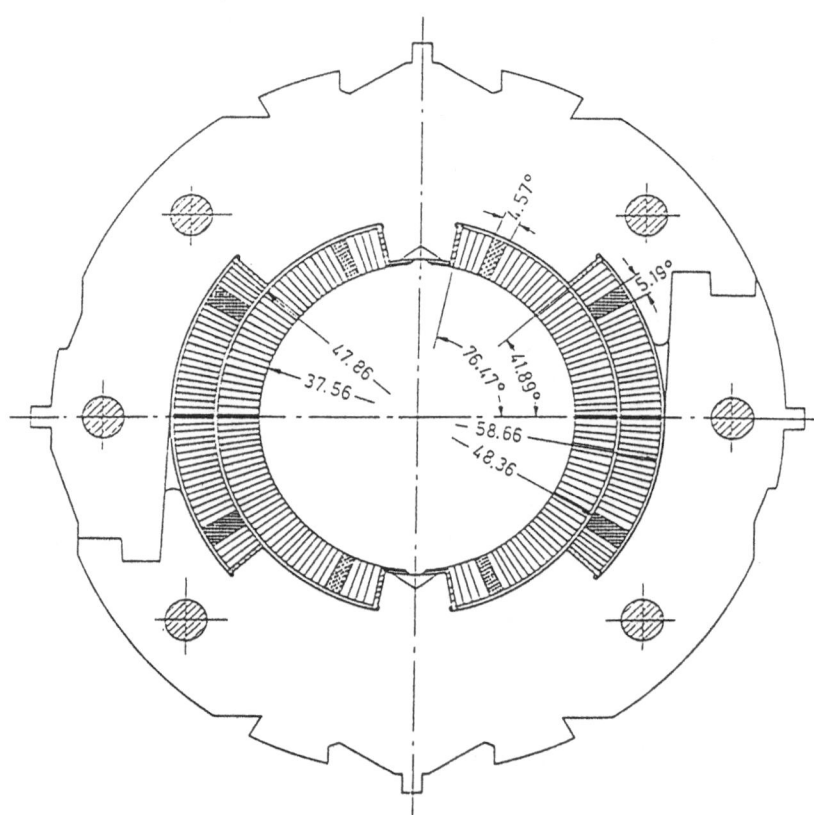

Fig. 1a.   Cross section of the HERA dipole coil.
           The coil is clamped with aluminum collars and surrounded by a cold
           iron yoke with an inner bore radius of 88.4 mm.

Cross sections of the HERA dipole[9] and quadrupole[10] are shown in Figs.1a,b. Both coil types are wound from a keystoned Rutherford-type cable with 24 strands in the dipole and 23 strands in the quadrupole. Half of the dipoles are produced in Italy with LMI superconductor, the other half in Germany with ABB conductor. The quadrupole production is split between France and Germany but all coils are made from Vacuumschmelze conductor. The parameters of the superconductors are summarized in Table 1.

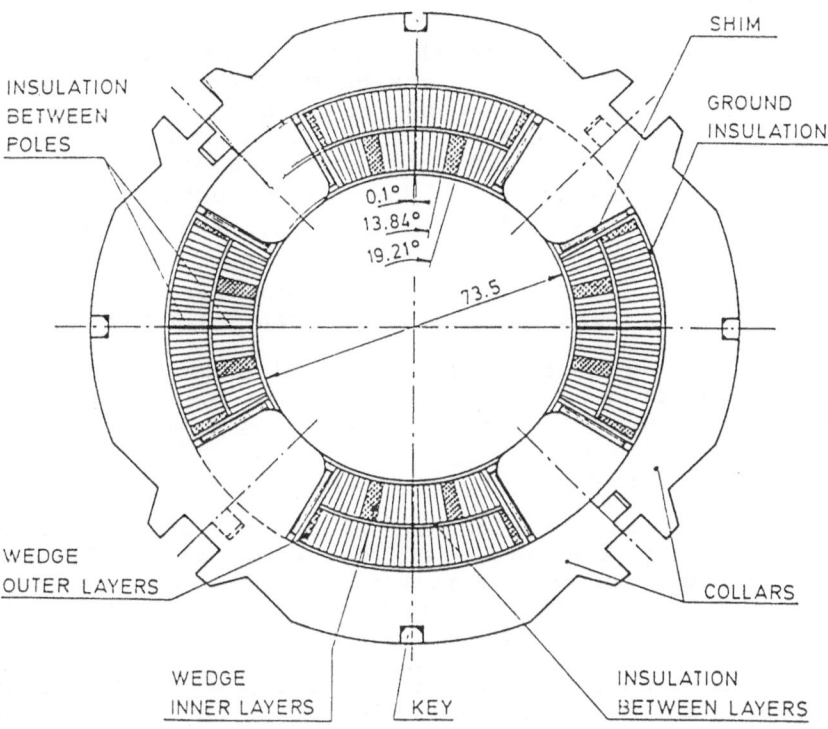

Fig. 1b.   Cross section of the HERA quadrupole coil.
The collars are made of stainless steel and are surrounded by a cold iron yoke with an inner radius of 78 mm.

The field of a superconducting accelerator magnet is conveniently expressed in terms of a multipole expansion (see e.g. Ref. 11). Taking the beam direction as the z axis of a cylindrical coordinate system (r, θ, z) the general expressions for the azimuthal and radial field components read

$$B_\theta\,(r,\theta) \;=\; B_{ref} \sum_{n=1}^{\infty} \left(\frac{r}{r_0}\right)^{n-1} \left(\; b_n \cos(n\theta) + a_n \sin(n\theta) \;\right)$$

$$(1)$$

$$B_r\,(r,\theta) \;=\; B_{ref} \sum_{n=1}^{\infty} \left(\frac{r}{r_0}\right)^{n-1} \left(- a_n \cos(n\theta) + b_n \sin(n\theta) \;\right)$$

Here r_0 is the reference radius of the expansion which is chosen to be 25 mm for the HERA magnets, i.e. 2/3 of the inner coil radius. The b_n and a_n are the normal and skew multipole coefficients. B_{ref} is a reference field which depends on the type of magnet. For dipole magnets, B_{ref} is chosen to be the dipole field itself whereas for quadrupoles the product of the gradient g and the reference radius r_0 is taken. With this convention, $b_1 = 1$ for an ideal dipole and all other coefficients vanish; similarly, for an ideal quadrupole $b_2 = 1$ and all other b_n and a_n are zero. Real magnets have of course non-vanishing higher multipole components due to mechanical imperfections, but the quality requirements are stringent for magnets to be used in a storage ring with beam lifetimes of many hours. For HERA the tolerance on the higher-order multipoles is $1 \cdot 10^{-4}$ except for the normal sextupole, which may be as large as $4 \cdot 10^{-4}$ The corresponding limits are $3 \cdot 10^{-4}$ for the quadrupole magnets which contribute less to the overall accelerator field since they are shorter, weaker and less numerous than the dipoles.

The HERA magnets have an inner coil diameter of 75 mm and are equipped with a cold beam pipe with an inner diameter of 55 mm. On the cryogenic test stands, a titanium tube of 45 mm diameter, surrounded by superinsulation, is mounted inside the evacuated beam pipe.

The multipole coefficients of the magnets are measured using rotating pick-up coils at room temperature. The induced voltage is digitized with a voltage-to-frequency converter and is summed in an up-down counter whose contents is read out at 128 equally spaced angular steps each revolution.

For the dipole magnets, pick-up coils with a length of 2.4 m are used. To cover the entire magnetic length of 8.28 m the measurements are performed in four longitudinal positions. In the following we use only the data from the first position where no sextupole and quadrupole correction coils are mounted on the beam pipe. These correction coils have a sizable influence on the persistent current fields, but this effect will be analysed elsewhere.

The quadrupole magnets are measured with a 1.1 m long pick-up coil. The total magnetic length of about 1.9 m is covered with two longitudinal positions. In this case the multipole data integrated over the full length can be used.

The pick-up coils for the measurements of dipole and quadrupole magnets are designed to cancel the induction resulting from the main poles which are typically four orders of magnitude larger than the disturbing higher-order poles. With this technique an accuracy of $1-2 \cdot 10^{-5}$ relative to the main poles is achieved.

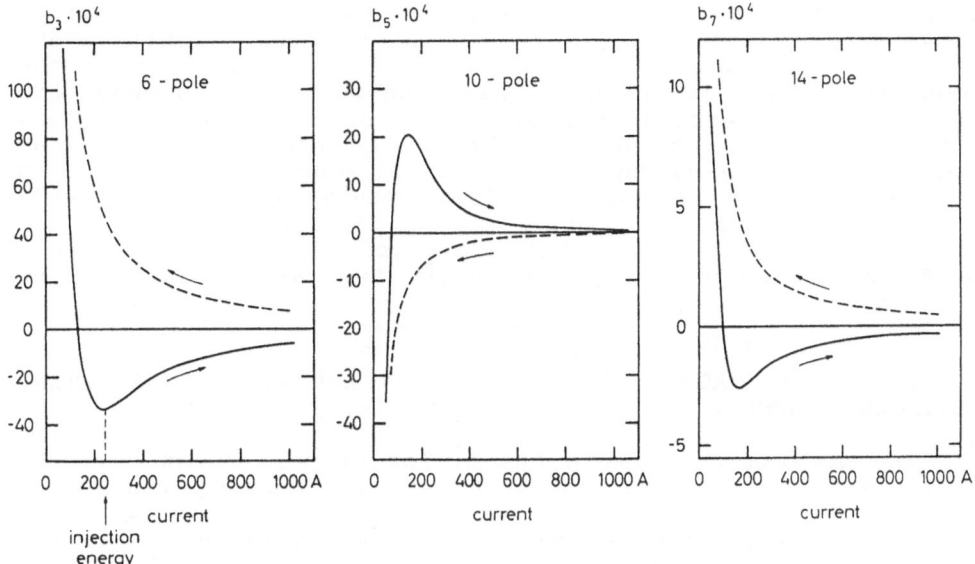

Fig. 2a. Current dependence of the 6-, 10- and 14-pole coefficients of a HERA dipole. Following an initial currend cycle O A -> 6000 A -> 50 A, the current is increased to 6000 A (continuous curves) and then decreased to 50 A again (dashed curves).

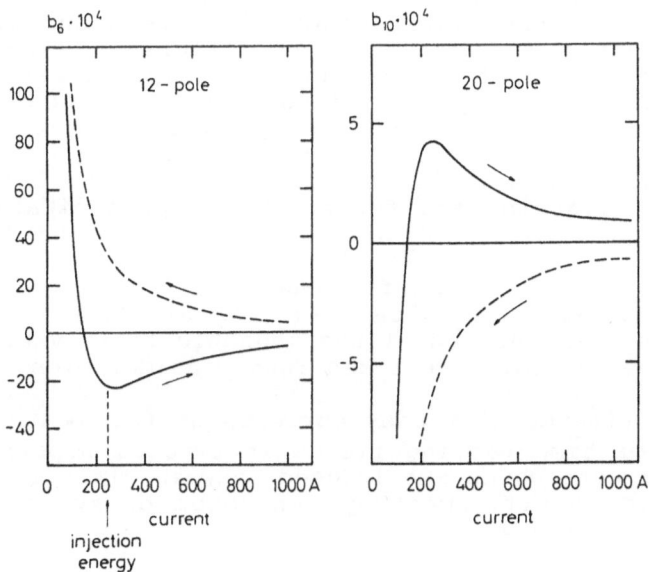

Fig. 2b. Current dependence of the 12- and 20-pole coefficients of a HERA quadrupole

The measured 6-, 10- and 14-pole coefficients of a HERA dipole and the 12- and 20-pole coefficients of a quadrupole are plotted in Figs. 2a,2b as a function of the magnet current. One observes a strong current dependence and a distinct hysteresis behaviour. For increasing currents the multipole coefficients follow the "up-ramp" curve while for decreasing currents the "down-ramp" curve is measured. Assuming that induced eddy currents in the superconductor are the source of the disturbing multipole fields, the hysteresis can be understood qualitatively since these currents change their sense of rotation when the ramp direction of the main field is reversed. For a quantitative understanding of this effect a detailed physical model is needed. In particular, the observation that the multipole coefficients change their sign on the "up-ramp" branch has to be explained.

CALCULATION OF MAGNETIZATION CURRENTS AND OF THE RESULTING MULTIPOLE FIELDS

The program to compute the fields of persistent currents in dipole and quadrupole magnets consists of three parts:

In the first step the main field, generated by the transport current in the coil, is calculated both within the useful aperture of the magnet and at any place inside the coil windings. A knowledge of this field is necessary since it is the time variation of just this "local" field which induces the eddy currents in the superconductor. The field calculation can be done analytically with good accuracy and will not be described here. Details can be found in Ref.11.

In the second step, the eddy current pattern is computed. It is an essential feature of twisted multifilamentary superconductors that long-lasting eddy currents occur only within single filaments and not between different filaments of a strand or between different strands of a transposed cable. Any change of the local magnetic field induces a new eddy current within an individual filament which is superimposed with the already existing persistent current. By closely following the history of the field it is therefore possible to calculate the eddy current distribution for an arbitrary excitation of the magnet.

In the last step, the field generated by a representative number of filaments is calculated. For simplicity, the vector potential is used instead of the magnetic field vector and the symmetries of the coil are taken into consideration.

In the treatment of eddy currents we follow M.N. Wilson's analysis[12] of time varying fields in type II superconductors. The eddy currents which are induced between different filaments of a twisted multifilamentary conductor decay exponentially with a time constant[12]

$$\tau = \frac{\mu_0}{2\rho} \left(\frac{L}{2\pi}\right)^2 \qquad (2)$$

Here ρ is the resistivity of the copper matrix and L the twist length. For the HERA conductors with L = 25 mm the estimated time is less than 0.1 s. In the Rutherford-type cables as used in our magnets also the eddy currents between different strands decay so rapidly that they have no influence on the multipole measurements[1]

Persistent eddy currents exist therefore only within single filaments, provided the filament spacing is large enough like in our cable that proximity-coupling effects[14] can be neglected. To calculate these currents we use the experimentally verified "critical state" model[15] proposed by Bean. According to this model, a type II superconductor tries to expel any external field change by generating a bipolar current distribution with the highest possible density, namely the critical current density J_c (B,T) at the given local field and temperature. Since the local field can be considered as homogeneous on the scale of a filament diameter (10-20 µm) the resulting current pattern in the filament can be computed analytically with good accuracy.

In the following, we distinguish between the "external" field B_e, i.e. the field which induces the eddy currents in the filament and is identical to the local field mentioned above, and the "internal" field B_i which is generated by those eddy currents. Suppose the external field is increased from zero to a small value B_e. The induced current has to follow a $\cos\theta$-like distribution to generate a homogeneous inner field B_i which just cancels the external field in the current-free region of the filament (see Fig.3a). This region can be approximated by an ellipse with large half axis, a = r_f (filament radius), small half axis b and eccentricity e = b/a. The field inside the ellipse, generated by the eddy currents, is found by simple integration

$$B_i = - \frac{2\mu_0 \, J_c \, a}{\pi} \left(1 - e \, \frac{\arcsin \sqrt{1-e^2}}{\sqrt{1-e^2}}\right) . \qquad (3)$$

From the condition $B_i = -B_e$ one determines the eccentricity e as a function of the external field B_e. The highest field which can be shielded from the interior of the filament is called the "penetrating" field B_p and is obtained for an ellipse shrunk to a line, i.e. e = 0.

$$B_p = \frac{2\mu_0 \, J_c \, a}{\pi} \qquad (4)$$

Fig.3b shows the currents in the "fully penetrated" filament.

The applied field may be raised to much larger values than B_p which is only about 0.13 T for the HERA conductor. In

---

[1] Eddy currents with a fast time dependence are discussed in a recent paper by D.ter Avest and L.J.M van de Klundert[13]

that case the same current pattern is obtained as in Fig.3b but the field inside the filament is no more zero.

If now the field is decreased again, eddy currents with opposite polarity are superimposed because the superconductor tries to avoid a change of the inner field. A more complicated current pattern arises as indicated in Fig.3c.

The eddy current loops are assumed to be closed at the coil ends. The effect of the short coil ends on the integrated multipole fields can be neglected.

The magnetization (magnetic moment per unit volume) of the $\cos\theta$-like current distribution shown in Fig.3a is easily computed

$$M = - \frac{4}{3\pi} \mu_0 \, J_c \, a \, (1 - e^2) \qquad (5)$$

The peak magnetization is obtained for the fully penetrated filament

$$M_p = |M|_{max} = \frac{4}{3\pi} \mu_0 \, J_c \, a \qquad (6)$$

Note that the quantity M_p is not constant but field dependent. It assumes its maximum value for $B_e = B_p$ and decreases proportional to the critical current density J_c (B_e ,T) when the external field is raised far beyond the penetrating field.

In the presence of a transport current, the equations (3), (4), (5) and (6) have to be modified by a factor ($1-J_t/J_c$ (B,T)) since the current density available to the eddy currents is then given by $J_c - J_t$ This correction factor is negligible near the injection field where the transport current density is two orders of magnitude lower than the critical current density but it becomes significant at high excitation of the magnet.

From the equations (3) and (5) one can compute the magnetization as a function of the external field. The result is plotted schematically in Fig.4. We observe a hysteresis behaviour with three different states:
Starting at the virgin state the magnetization follows an initial curve (i) and reaches its peak value at $B_e = B_p$. After going up to high fields the ramp direction is reversed and M follows the "down-ramp" branch (d). At a certain minimum current (typically $I_{min} = 50$ A) the field is increased again and the magnetization follows the "up-ramp" branch (u) which has the remarkable feature that M changes its sign from positive to negative values. This is exactly what is observed in the sextupole and dodecapole coefficients (see Figs.2a,b). Fig.4 also indicates the current pattern in the filament at different positions of the hysteresis loop.

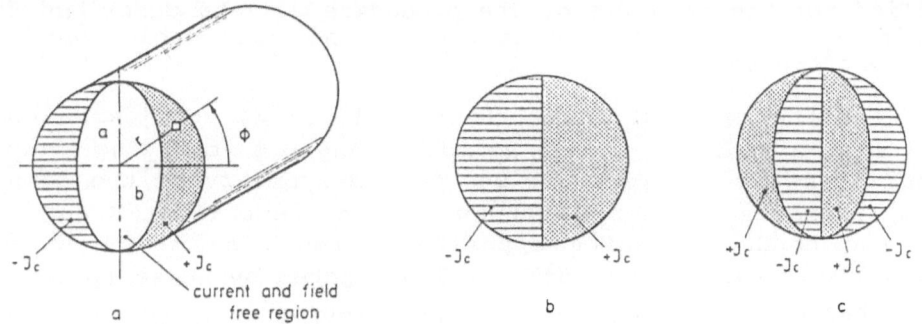

Fig. 3. Schematic view of the persistent eddy currents which are induced
in a superconducting filament by a time dependent external field.

a) The external field in raised from zero to a value B_e less than
the penetrating field B_p.

b) A "fully penetrated" filament, i.e. $B_e \geq B_p$.

c) Current distribution which results when the external field is
first increased from zero to a value above B_p and then decreased
again.

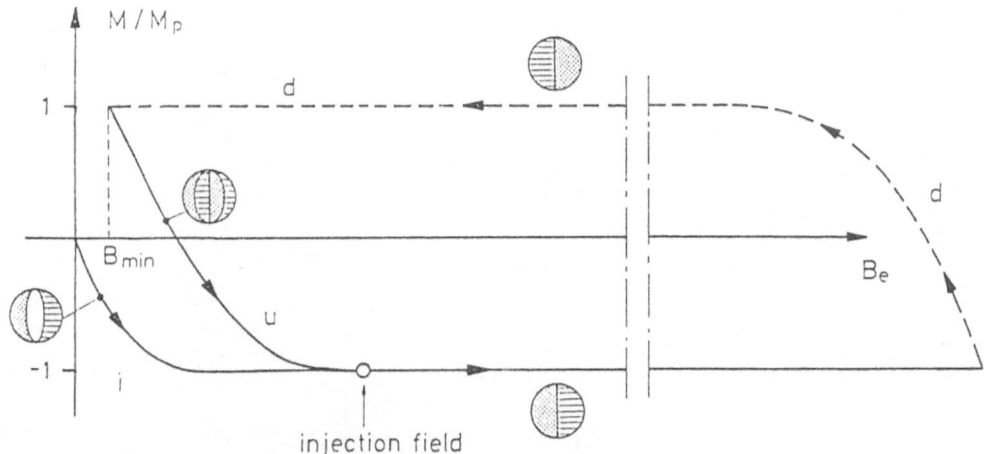

Fig. 4. The normalized magnetization M/M_p of a NbTi filament as a function
of the external field. (i): initial curve, (u): up-ramp branch,
(d): down-ramp branch. Also shown are the current distributions in
the filament.

The calculation of the multipole fields caused by the
persistent currents can be performed with similar methods as

applied for the main field. The procedure will be described for a dipole only, a generalization to the quadrupole case is straightforward.

For the computation of the main field we consider a conductor at a radius R and an azimuthal angle ϕ, carrying a current +I in the z direction. Due to the symmetry of the dipole coil, there is another conductor with current +I at R and $-\phi$ and two conductors with opposite currents at radius R and angles $\pi \pm \phi$. The vector potential generated by these four currents has only a z component and is given by[11]

$$A_1 (r, \theta) \quad \frac{2\mu_0 I}{\pi} \sum_{n=1,3,\ldots} \frac{1}{n} \left(\frac{r}{R}\right)^n \cos (n\theta) \cos (n\phi) \tag{7}$$

Only odd multipole orders n = 1, 3, 5,..appear in the expansion.

To generalize this expression for the case of eddy currents we replace the current distribution in each filament by a pair of line currents +I and −I whose strength equals the integrated current density and whose separation d is chosen such that the computed filament magnetization (5) is obtained. (d is about equal to the filament radius). Fig.5a illustrates that the eddy currents obey the same symmetry as the transport current.

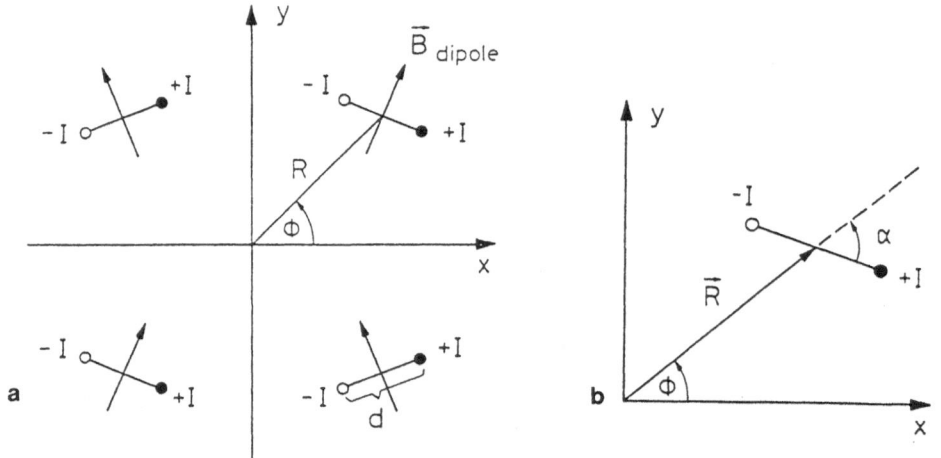

Fig. 5. a) Eddy currents induced by the time-dependent main field in four symmetrically arranged filaments inside the dipole coil. The separation d between the positive and negative currents is grossly exaggerated.
b) Definition of the angle α between the line joining the current pair and the position vector R.

For d « R, the vector potential of the four current pairs in Fig.5a is given by

$$A_2 = \frac{\partial A_1}{\partial R} \Delta R + \frac{\partial A_1}{\partial \phi} \Delta \phi$$

With the relations $\Delta R = d \cdot \cos\alpha$, $\Delta\phi = -d \cdot \sin\alpha/R$ (see Fig. 5b for the definition of the angle α) we obtain

$$A_2(r,\theta) = \frac{2\mu_0 I}{\pi} \sum_{n=1,3,\ldots} \left(\frac{r}{R}\right)^n \cos(n\theta) \cos(n\phi+\alpha) \tag{8}$$

The influence of the iron yoke with an inner bore radius R_I is taken into account by the image current method (see e.g. Ref.11). The image of a current pair at a radius R and angle ϕ appears at $R' = R_I^2/R$ and $\phi' = \phi$. The separation of the image currents is $d' = d \cdot R'/R$ and the angle with respect to the position vector R' is $\alpha' = \pi - \alpha$. Replacing the quantities R, d, α in equ.(8) by R', d' and α' one gets the iron contribution A'_2 to the vector potential. The resulting multipole expansion of the azimuthal field component is then given by

$$B_\theta(r,\theta) = -\frac{\partial}{\partial r}\left(A_2 + A'_2\right) \tag{9}$$

$$= \frac{2(\mu_0 I d)}{\pi R^2} \sum_{n=1,3,\ldots} n \cos(n\theta) \left[\left(\frac{r}{R}\right)^{n-1}\cos(n\phi+\alpha) - \frac{R}{R'}\left(\frac{r}{R'}\right)^{n-1}\cos(n\phi-\alpha)\right]$$

For the product $(\mu_0 I d)$ we insert the magnetic moment per unit length derived from equ.(5). Expression (9) has to be summed over all NbTi filaments in one quarter of the dipole coil. Sufficient accuracy is obtained by using one filament for each strand in performing the summation and by multiplying the result with the number of filaments per strand. The multipole coefficients generated by the persistent eddy currents are finally

$$b_n = \left(\sum_{\text{filaments}} B_{\theta,n}(r=r_0, \theta=0)\right)/B_0 \tag{10}$$

Here B_0 is the dipole field of the transport current. As for the main field only normal multipoles of the orders n = 1,3,5,.. occur.

In the quadrupole coil, the persistent eddy currents obey again the same symmetries as the transport current and generate therefore only normal multipoles whose order is an odd multiple of the basic pole order, i.e. the only nonvanishing poles are b_2, b_6, b_{10}, b_{14}, ...

An important ingredient to the model is the critical current density J_c (B,T) at low fields. Unfortunately, critical current data at low fields are not easily accessible. The manufacturers of superconducting cables measure critical currents usually at fields of 5-6 T. No direct measurements exist at the fields of interest. A few superconductor magnetization measurements have been performed by Ghosh and Sampson[16] at BNL on HERA conductors. From these one can derive the critical current density by making use of expression (6) and correcting for the volume fraction of superconductor in the cable.

We have evaluated the BNL measurements on the ABB dipole conductor and have parametrized the critical current density in the following form

$$J_c \quad = 2.3 \cdot 10^{10} \text{ A/m}^2 \text{ for B} \leq 0.025 \text{ T}$$

$$J_c \text{ (B)} = 1.7 \cdot 10^{10} \left(\frac{B}{0.1T}\right)^{-0.2} \text{A/m}^2 \text{ for } 0.025 \text{ T} < B \leq 0.25 \text{ T}$$

$$J_c \text{ (B)} = 1.5 \cdot 10^{10} \left(\frac{B}{0.25T}\right)^{-0.5} \text{A/m}^2 \text{ for B} > 0.25 \text{ T}$$

The numbers refer to a helium temperature of 4.6 K.

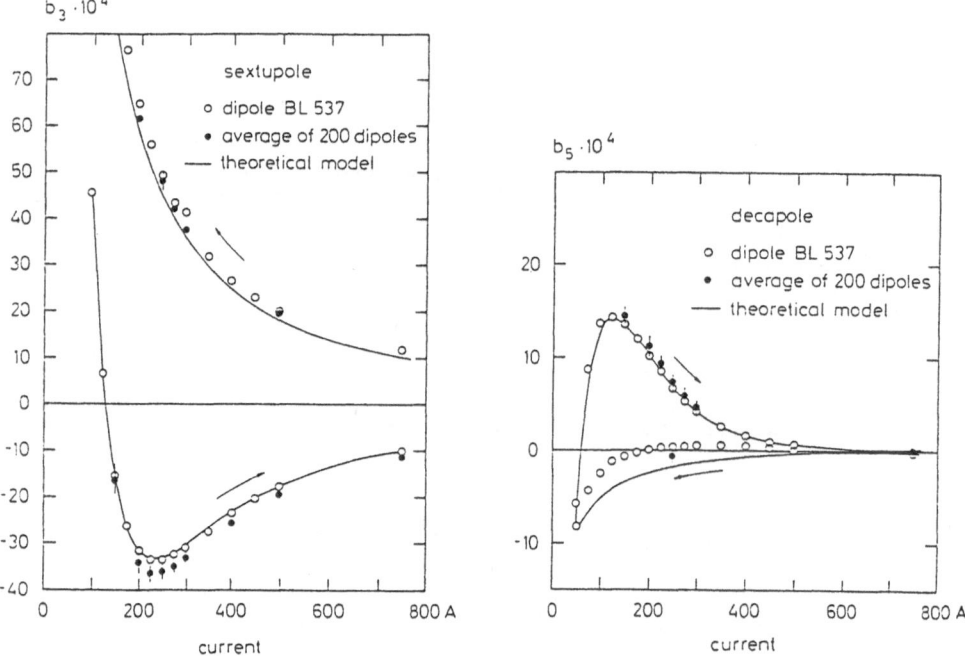

Fig. 6. Measured multipole coefficients of a single dipole and the average coefficients of 200 dipoles plotted against the current in the coil.
The magnets were cooled by single-phase helium in forced flow with a temperature of about 4.6 K.
Continuous curves: predictions of the eddy current model.

We estimate an uncertainty of 10-20 % in our knowledge of critical current data at low fields. This is at present the largest source of uncertainty in the calculation of persistent current multipoles. The filament diameter is known to about 5%.

The multipole data measured for a selected dipole magnet (serial number BL537) at a large number of current values between 50 A and 750 A are plotted in Fig.6. The sextupole and decapole coefficients show a smooth current dependence on both hysteresis branches. Very good reproducibility ($2 \cdot 10^{-5}$) is found when the measurement is repeated. Due to time limitations the majority of the magnets are measured only at a few selected currents. The average data of 200 dipoles are also plotted in Fig. 6, together with their rms standard deviations. These data have been collected over a period of 16 months and agree very well with the data of dipole BL537.

Also shown in Fig.6 are the predictions of the eddy current model. The agreement with the data both in curve shape and absolute magnitude is excellent considering the uncertainties in the critical current density mentioned above. Only on the down-ramp branch of the decapole coefficient there is some discrepancy between the measured and calculated values. During the accelerator operation the up-ramp branch will be used.

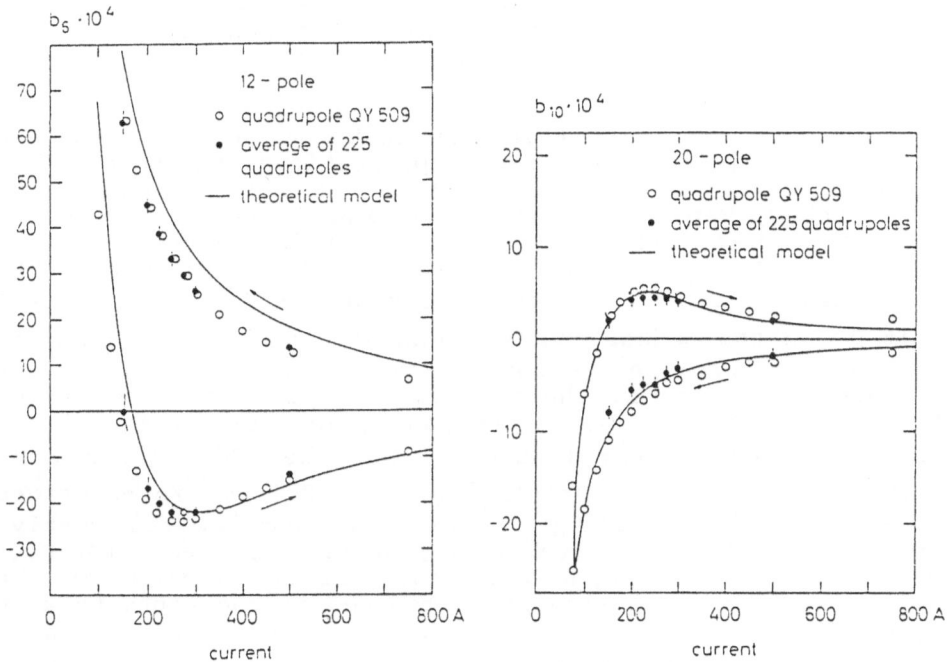

Fig. 7.   Measured multipole coefficients of a single quadrupole and the average coefficients of 225 quadrupoles. Continuous curves: predictions of the eddy current model.

In Fig.7 we present the multipole data of a single quadrupole (QY 509) and the average data of 225 quadrupoles. Again the measurements reveal good internal consistency and reproducibility. The theoretical curves provide a good description also in this case, but somewhat larger differences are observed on the down-ramp branch of the 12-pole coefficient.

Due to the lack of critical current data for the quadrupole conductor we have used the parametrization made for the ABB conductor which probably overestimates the current density in the VAC conductor by about 5 %.

We want to emphasize that the theoretical curves shown are absolute predictions and no parameter has been adjusted to yield a good fit to the data.

The persistent currents have also a significant influence on the main field of the dipoles and on the quadrupole gradient. Their contribution is denoted by $\tilde{B}_1$ resp. $\tilde{g}$ and is determined by taking the difference between the measured and computed values. The latter ones are derived from measurements at higher currents where the persistent current effects are small and the yoke saturation still negligible. The data for $\tilde{B}_1/B_1$ and $\tilde{g}/g$ are plotted in Fig.8. Again a hysteresis curve is observed and the data are in good agreement with the model prediction. (It should be mentioned that the dipole field contains in addition a small contribution from the remanence of the iron yoke.) At the HERA injection energy the main field components are 0.3 - 0.5% lower than the values computed from the coil current. Of course a correction is needed to match HERA to the energy of the pre-accelerator.

Equations (5), (9), (10) show that the magnetization and all higher multipoles are directly proportional to the critical current density at low field and the filament diameter. Of course, nobody wants to sacrifice a high J just to reduce the persistent current effects but a reduction in filament diameter is certainly advisable. The superconducting cable in the HERA dipoles has a fairly large filament diameter (14 μm in the ABB and 16 μm in the ANSALDO-ZANON magnets). When the HERA superconductor was specified, a not too small filament diameter was considered important by the manufacturers to guarantee a high critical current density. In the past years great progress has been made towards finer filaments. For the SSC magnets 6 μm diameter is foreseen at present but even 2.5 μm are envisaged if the production should prove reliable and not too costly. There is an interesting lower limit, however, at least for NbTi embedded in copper. With decreasing filament diameter the filament spacing w has also to be reduced if one wants to keep a constant copper to superconductor ratio. For w below 1 μm a "proximity coupling" between neighbouring filaments has been observed, basically a tunneling of the Cooper pairs through the copper.

Figure 9 shows the computed sextupole coefficient[14] in an SSC dipole at a field of 0.33 T, plotted against the filament diameter d with the ratio w/d as parameter. For w/d = 0.2 the

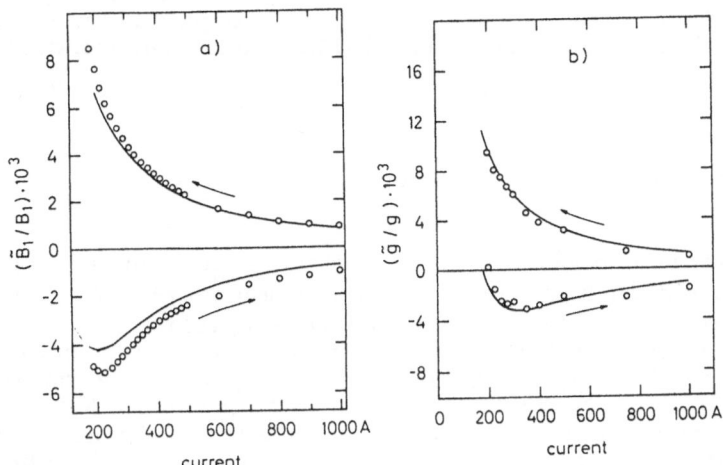

Fig. 8. The contribution $\tilde{B}_1$ ($\tilde{g}$) of the persistent currents to the main dipole field (main quadrupole gradient). Solid curves: model prediction.

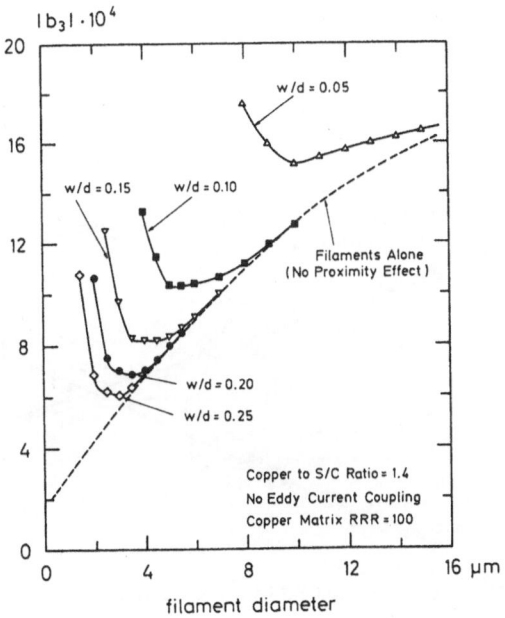

Fig. 9. Calculated sextupole in an SSC dipole as a function of the NbTi filament diameter; w is the interfilament spacing.

optimum filament diameter is about 4 μm. Any further reduction leads to a steep increase in the multipole fields.

The proximity coupling might be inhibited by a resistive barrier between the filament and the copper matrix but this would certainly have a negative influence on the conductor stabilization.

TIME DEPENDENCE OF PERSISTENT CURRENT EFFECTS

A time dependence of the persistent current sextupole was first observed in the Tevatron dipole magnets[17]. From measurements on preseries HERA magnets we were able to show[11] that the multipole fields decrease proportional to the logarithm of time, contrary to the exponential decay of eddy currents in circuits with inductive and resistive components. Flux creep in hard superconductors is known[18] to lead to a logarithmic time decrease of the critical current density, and this phenomenon is the presently accepted explanation of the time dependence of the multipole fields.

Figure 10 shows the historic measurements of Kim et al.[18] on the time variation of the trapped or expelled magnetic flux in a superconducting ring which follows a log t law for times between 10 s and 5000 s. A similar behaviour has been found in the HERA magnets although here the flux is not trapped in superconducting cylinders but rather in the fine filaments of a Rutherford type cable.

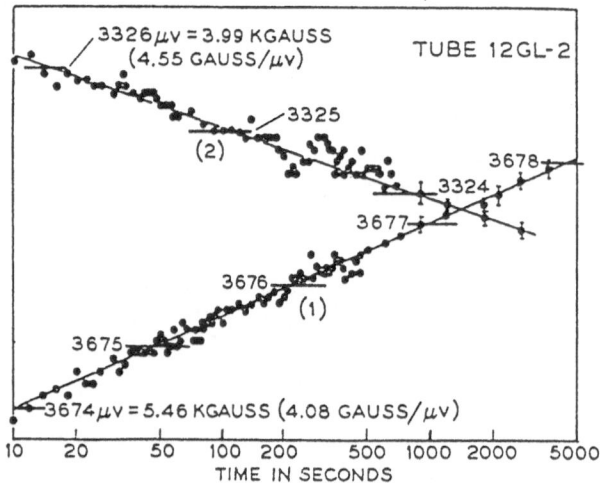

Fig. 10. Decay rate measurements of "persistent" currents induced in a 3Nb-Zr tube (from Ref. 18). The data in (1) were taken at a point on the shielding portion of the magnetization curve and (2) on the trapping portion.

In Fig.11 the absolute value of the sextupole coefficient in a HERA dipole is plotted against the logarithm of time. As time origin we have chosen the instant at which the coil current reached the value of 250 A. The sextupole coefficient decreases by about 25% within 10 hours. Between 200 s and 4000 s the drop is almost linear; at larger times but also below 100 s the slope levels off.

The absolute value of the persistent current contribution $\tilde{B}_1$ to the main dipole field (Fig. 12) has a similar logarithmic decrease. Also the time variation of the 12-pole in a quadrupole (Fig. 13) resembles that of the sextupole.

The resolution of our "stretched-wire" system[1] is barely sufficient to determine the time dependence of the quadrupole gradient g. The relative change between 200 s and 2000 s, averaged over 3 quadrupoles, is $\Delta g/g = (1.2\pm0.6)\cdot10^{-4}$ compared to an estimated $1.5\cdot10^{-4}$ from the observed time variation of the 12-pole.

To perform such time dependence measurements well-defined initial conditions are needed. The magnet is quenched first and then the current is cycled 50 A - 6000 A - 50 A at a rate of 10 - 20 A/s. The minimum current of 50 A has been chosen to ensure a proper power supply regulation, which was found to be essential for obtaining reproducible results. The current of 250 A (corresponding to the injection energy) is usually approached with a low speed of 1 A/s so as to reduce the current overshoot to less than 0.3 A.

The injection of protons into the HERA ring is estimated to take about 30 minutes. Therefore the change of the field components is determined for times between 200 s and 2000 s. Here the data are well represented by the form A-R·lg(t) so this change is identical to the logarithmic decay rate R.

The decay rates of the sextupole in 51 dipoles and of the 12-pole in 88 quadrupoles are presented in Fig. 14. The first observation is that the distributions are rather wide. Secondly, within the limited statistics, the average decay in the dipoles made with LMI superconductor appears to be twice as large ($(3.7\pm0.6)\cdot10^{-4}$) as in the dipoles with ABB superconductor ($(1.8\pm0.5)\cdot10^{-4}$). That a difference exists may not be too surprising since the rate of flux creep depends on the properties of the pinning centers and these are influenced by the wire production process. Unfortunately, no quantitative theory is available and even the parameters that influence the decay rate of the critical current density are unknown. Experimental uncertainties like insufficient power supply regulation may have contributed to a broadening of the distributions. However, for a single magnet which has been measured repeatedly over a period of 2 months, a narrow distribution has been obtained (see insert to Fig.14a).

In Fig.15 we have plotted the logarithmic decay rate of the sextupole against that of the dipole. A very clear correlation is observed, supporting the hypothesis that the variation of these quantities is caused by a time dependence of the critical current density. Note that different experimental methods

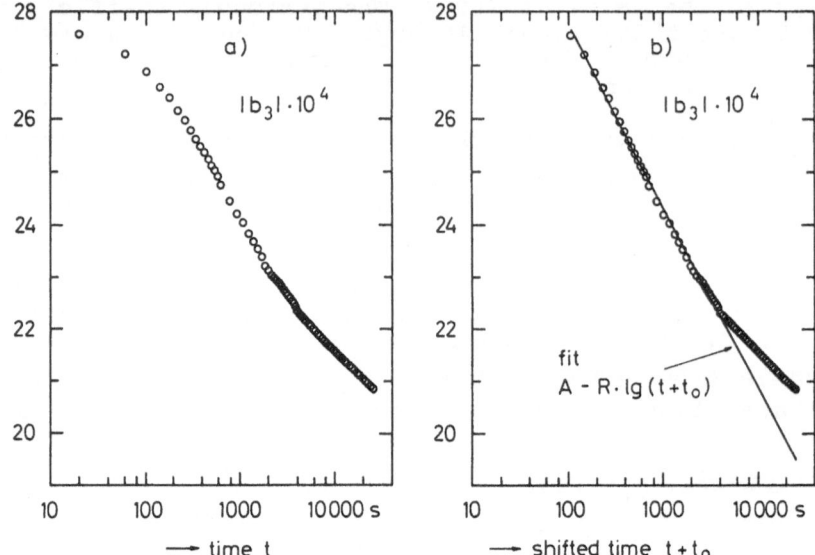

Fig. 11. a) Time dependence of the absolute value of the sextupole in a HERA dipole. b) Same data plotted on a shifted time scale $t + t_o$.

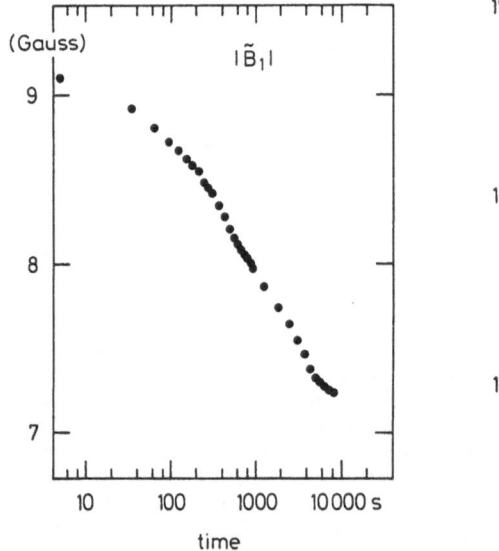

Fig. 12. Time dependence of the eddy current contribution $\tilde{B}_1$ to the main dipole field.

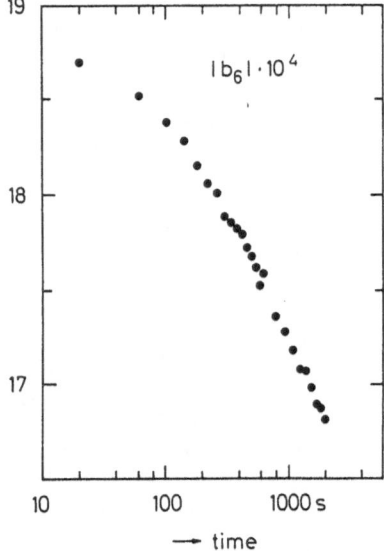

Fig. 13. Time dependence of the 12-pole in a quadrupole.

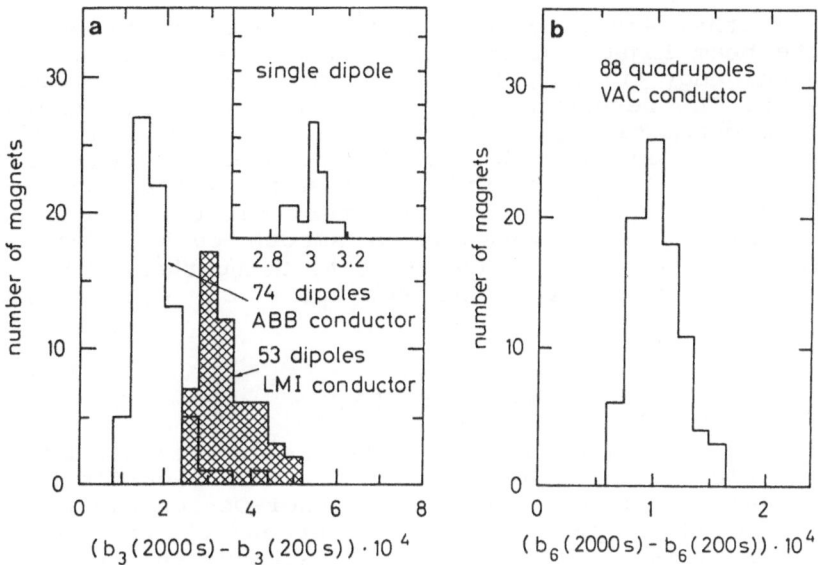

Fig. 14. Decay rates (change between 200s and 2000s) of a) sextupole in
dipoles and b) 12-pole in quadrupoles.

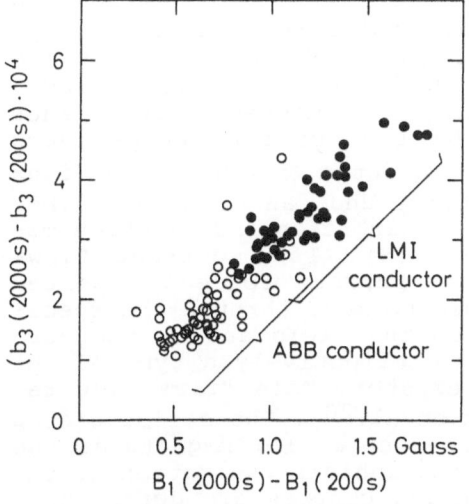

Fig. 15. Correlation between the
decay rates of the sextupole
and of the main field in the
dipole magnets.

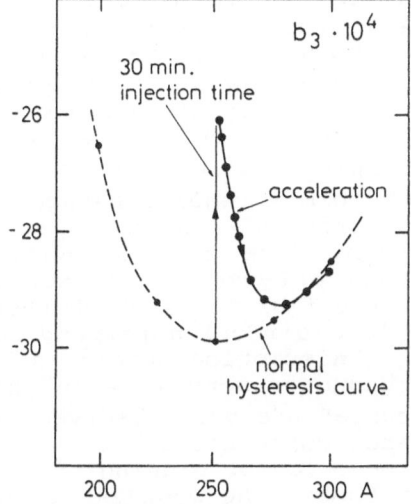

Fig. 16. Time variation of the sextupole coefficient in a dipole during the injection
and acceleration phase in
HERA.

are used to determine the two quantities: the dipole is measured with an NMR probe, the sextupole with a rotating pickup coil.

A further study concerns the acceleration phase in HERA. When the beam injection has been finished and the acceleration starts new eddy currents will be induced in the superconductor filaments. The resulting sextupole changes in the dipoles may shift the chromaticity, as observed in the Tevatron. In Fig.16 we have simulated the injection and acceleration phase. During the 30 minute waiting period at 250 A the sextupole b_3 drifts away from the hysteresis curve. After that the dipole current is increased and b_3 moves on a smooth curve, joining the hysteresis curve again at about 280 A. Using these data it will be possible to program the sextupole correction currents appropriately.

SUPERCONDUCTORS IN A MAGNETIC FIELD

It has been known since the famous experiments by Meißner and Ochsenfeld that superconductors show a peculiar behaviour in external magnetic fields. The Meißner-Ochsenfeld effect implies that the interior of a superconducting specimen is shielded from magnetic fields by surface currents provided the field strength stays below a critical value B_c (T). The genuine "Meißner" effect is observed in type I superconductors which are in general pure elements like lead or mercury. A wide range of superconductors, mostly alloys but also the important superconducting element niobium, show a more complicated behaviour. These so-called "type II" superconductors are characterized by two critical fields, B_{c1}(T) and B_{c2}(T). The upper critical field B_{c2}(T) may exceed by far the critical field B_c(T) of a type I superconductor. For magnetic fields below B_{c1}(T) a type II superconductor stays in the Meißner phase with complete field exclusion. For fields between B_{c1} (T), and B_{c2}(T), however, the material enters the mixed phase: magnetic field lines penetrate the bulk material in the form of flux tubes, each containing a single elementary flux quantum $\phi_0 = h/2e$. The flux tubes form a regular hexagonal lattice and can be made visible by decoration with fine paramagnetic particles and electron microscopy (see e.g. W. Buckel, Ref.19). If a current flows through a type II superconductor a Lorentz force is exerted upon the flux tubes which then begin to move through the specimen in a direction perpendicular to the field and to the current. This motion generates heat, so effectively a type II superconductor acts like an ohmic resistor. This "flux flow resistance" has been observed by Kim et al.[20] . The decisive step towards constructing resistance-free cables for magnets is the introduction of pinning centres which inhibit the motion of the flux tubes. The most important pinning centres are normal precipitates and lattice defects. The properties of the pinning centres depend strongly on the various cold drawing and heat treatment steps during the wire production and here is where the art of making a good superconductor comes in. A so-called "hard" superconductor with good pinning centres exhibits no flux flow and therefore no apparent ohmic resistance when currents are flowing in the presence of a magnetic field.

Nevertheless, there are still some "flux creep" effects

left: at nonvanishing temperatures (even at 4.2 K) the thermal energy is sufficient to release a few flux tubes from their pinning centres which then move through the specimen. During their motion they generate an apparent resistance which reduces the critical current density. In his explanation of the Kim experiment[17], Anderson[21] has proposed a simple model for the time dependence of persistent currents. The pinning centres are represented by potential wells of depth U_0 and width b, in which bundles of flux quanta are captured. The "hopping" rate is given by an expression

$$R_0 = v_0 \ \exp(-U_0 \ /kt)$$

where v_0 is the vibration frequency of the bundle inside the well and kT the thermal energy. If a current with density J is flowing in the conductor, a Lorentz force arises

$$F = V \ J \cdot (n\phi_0).$$

Here V is the volume of the probe and n the number of flux quanta in the well. The potential acquires now a slope and the effective well depth is reduced by

$$\Delta U = F \cdot b = V \cdot J \cdot n\phi_0 \ b$$

Then the hopping rate becomes

$$R = v_0 \ \exp(-(U_0 - \Delta U)/kT) = R_0 \ \exp((VJn\phi_0 \ b)/kT).$$

Consider now the case of a superconducting cylinder with a trapped magnetic field. The current density in the cylinder is proportional to the trapped field. So the hopping rate is

$$R = R_0 \ \exp(B/B_1)$$

where the quantity B_1 contains the other factors in the exponential. If a magnetic flux bundle is released from its pinning centre it will be moved out of the cylinder by the Lorentz force, thereby reducing the trapped magnetic field. So we arrive at an approximate equation

$$\frac{dB}{dt} = C \ \exp\left(B/B_1\right)$$

(C is roughly a constant) which has the solution

$$B(t) = B_0 - B_1 \ \ln(1 + t/t_0).$$

Although a lot of approximations have been used in this derivation the resulting log t - behaviour seems to be quite universally valid.

The critical current density is not really a well-defined quantity since some logarithmic decay is observed. For a practical definition of the critical current density we may choose the value which is established after the decay on a linear time scale has become unmeasurably small.

Table 1. Parameters of the superconducting cables

	ABB dipole cable	LMI dipole cable	VAC quadrupole cable
strand diameter/mm	0.84	0.84	0.84
filament number	1230	900	636
filament diameter/μm	14	16	19
Cu: NbTi ratio	1.8	1.8	1.8
twist pitch/mm	25	25	25
number of strands per cable	24	24	23
cable pitch/mm	95	95	95
average critical current at 5.5 T, 4.6 K	8870 ± 220 A	8320 ± 330 A	8030 ± 90 A

REFERENCES

1. H.R. Barton Jr., R. Bouchard, Yanfang Bi, H. Brück, M. Dabrowska, D. Darvill, Wenliang He, Zhuomin Chen, Zhengkuan Jiao, D. Gall, G. Knies, J. Krzywinski, J. Kulka, A. Ladage, R. Lange, Liangzhen Lin, A. Makulski, R. Meinke, F. Müller, K. Nesteruk, J. Nogiec, H.Preissner, W.Rakoczy, P.Schmüser, M.Surala, W.Schnacke, Z.Skotniczny, Wenlong Shi, Contribution to the 11th International Conference on Magnet Technology MT-11, Tsukuba, Japan 1989 and DESY report HERA 89-20 (1989)
2. H. Brück, Zhengkuan Jiao, D. Gall, G. Knies, J.Krzywinski, R. Meinke, H. Preissner, P. Schmüser, Contribution to the 11th International Conference on Magnet Technology MT-11, Tsukuba, Japan 1989 and DESY report HERA 90-01 (1990)
3. For a recent review of the HERA collider see B. H. Wiik, Proceedings of the XXIV International Conference on High Energy Physics, München 1988
4. C. Daum, J. Geerinck, P. Schmüser, R. Heller, H. Möller, M. Muto, S. Schollmeier, P.A.M. Bracke, IEEE Trans. MAG. 24, No.2, 1377 (1988) and DESY report HERA 89-09 (1989)
5. R. Brinkmann and F. Willeke, DESY report HERA 88-08 (1988)
6. M. A. Green, IEEE Trans., NS-18, No.3 (1971) 664
7. J.-L. Duchateau, Departement SATURNE internal report SEDAP/72-109 (1972)
8. H. Brück, R. Meinke, F. Müller, P. Schmüser, Z. Phys. C-Particles and Fields 44, 385 (1989)

9.   S. Wolff, DESY report 87-116 (1987); P. Schmüser, Proceedings of the Oregon APS Meeting, August 1985

10.   A. Auzolle, P. Le Marrec, A. Patoux, J. Perot and J.M. Rifflet, Contribution to the ICFA Workshop on Superconducting Magnets and Cryogenics, Brookhaven, May 1986, BNL report 52006

11.   K.-H. Mess and P. Schmüser, DESY report HERA 89-01 (1989) and Proceedings of the Course on Superconductivity in Particle Accelerators, Hamburg, May 30 to June 3, 1988.CERN yellow report 89-04 (1989)

12.   M. N. Wilson, Superconducting Magnets, Clarendon Press, Oxford 1983

13.   D. ter Avest and L.J.M. van de Klundert, University of Twente internal report, 1989

14.   A.K. Ghosh, W.B. Sampson, E. Gregory, T.S. Kreilick, IEEE Trans. Mag-23, No.2, 1724 (1987) M.A. Green, LBL-report 23823 (1987)

15.   C. P. Bean, Phys. Rev. Letters $\underline{8}$ (1962), 250; Y. B. Kim, C.F. Hempstead and A.R. Strnad, Phys. Rev. Letters $\underline{9}$ (1963), 306

16.   A.K. Ghosh and W.B. Sampson, private communication, see also A.K. Ghosh, K.E. Robins and W.B. Sampson, IEEE Trans. on Magnetics, Vol. 21 (1985), 328

17.   D.A. Herrup, M.J. Syphers, D.E. Johnson, R.P. Johnson, A.V. Tollestrup, R.W. Hanft, B.C. Brown, M.J. Lamm, M. Kuchnir, A.D. McInturff, IEEE Trans. Mag. 25, No. 2, 1643 (1989); R.W. Hanft, B.C. Brown, D.A. Herrup, M.J. Lamm, A.D. McInturff, M.J. Syphers, IEEE Trans. Mag.25, No. 2, 1647 (1989)

18.   Y.B. Kim, C.F. Hempstead and A.R. Strnad, Phys. Rev. Letters 9, 306, (1962)

19.   W. Buckel, Supraleitung, Physik-Verlag Weinheim, dritte Auflage 1984

20.   Y.B. Kim, C.F. Hempstead and A.R. Strnad, Phys. Rev. 139, A1163 (1965)

21.   P.W. Anderson, Phys. Rev. Letters 9, 309 (1962)

CORRECTION MAGNETS

C. Daum

NIKHEF-H

P.O. Box 41882, 1009DB Amsterdam, Netherlands

INTRODUCTION

Main elements of circular accelerators/colliders with separated function lattice are:

dipole magnets, which provide the guidance field;

quadrupole magnets, which provide the focusing of the beam particles around the mean beam orbit.

Using a cylindrical coordinate system with z along the beam axis, r the distance to the mean beam orbit and θ the angle with respect to the plane of the mean beam orbit, the components of the fields due to these magnets are given by the multipole expansion:

$$B_\theta(r,\theta) = B_0 \sum_{n=1}^{\infty} \left(\frac{r}{r_0}\right)^{n-1} \left(a_n \sin(n\theta) + b_n \cos(n\theta)\right) \delta_{n,(2k+1)p},$$

$$B_r(r,\theta) = B_0 \sum_{n=1}^{\infty} \left(\frac{r}{r_0}\right)^{n-1} \left(- a_n \cos(n\theta) + b_n \sin(n\theta)\right) \delta_{n,(2k+1)p},$$

where p is the basic multipole order of the 2p-pole magnet, k = 0,1,2,.., r_0 is the reference radius, and B_0 is the field at the reference radius. Here, a_n and b_n are the "normal" and "skew" multipole components of the field.

For the main dipole magnet p = 1, a_n = 0, and the non-zero normal components are

b_1, dipole,
b_3, sextupole,
b_5, decapole, etc.,

and the reference field is B_0, the strength of the main dipole field.

New Techniques for Future Accelerators III
Edited by G. Torelli, Plenum Press, New York, 1990

For the main quadrupole magnet $p = 2$, $a_n = 0$, and the non-zero normal components are

b_2, quadrupole,
b_6, dodecapole, etc.,

and the reference field is $B_0 = G\, r_0$, where G is the gradient of the field.

We choose the mean beam orbit in the horizontal plane, and the main dipole field is vertical to this plane.

CORRECTION ELEMENTS

The description is made for hadron machines with superconducting magnets. However, apart from superconductivity, electron machines have the same requirements.

In general, the beam optics requires that higher multipoles are less than 2×10^{-4} of the main field B_0, which is usually expressed as 2 units (of 10^{-4} of B_0).

Various correction magnets have to be added to the main structure for compensation of errors.

Orbit correction

Alignment errors of the main magnets with respect to the mean beam orbit, e.g. quadrupole axis not coinciding with mean beam orbit, azimuthal tilt of dipoles, etc.; correction elements needed:

horizontal and vertical dipoles in each cell of the machine lattice; horizontal(vertical) dipoles should be located near horizontally(vertically) focusing main quadrupoles.

Approximation of the field shape

The approximation of the main field by the shape of the coils of the main magnets; in practice, pure harmonic coils cannot be made, but are approximated by current shells with specific choice of angular range and having spacers for suppression of higher multipoles; also coil heads contribute to higher multipoles, which can be minimized by a judicious choice of the shape of the coil heads; in general, for a proper design of a magnet, no corrections are necessary.

Mechanical inaccuracies

These may be due to:

1) magnetic length, corrections needed are:
   for main dipoles: horizontal dipole corrections (as for orbit correction);
   for main quadrupoles: quadrupole corrections;

2) coil imperfections; with proper mechanical tolerances no corrections are needed.

Saturation effects of the iron yoke

The image currents add to the main pole and the higher multipoles, which are not linear with the excitation; correction elements needed are:

for main dipole: sextupole and decapole corrections;
for main quadrupole: dodecapole corrections.

Magnetization effects of superconducting cables

The persistent currents induced in the superconducting filaments of the conductors of the main magnets give rise to large multipoles, which are present over the whole range of excitation of the main magnet, but limit in particular the dynamic aperture of the machine at low main field levels at injection; corrections needed are:

for main dipole: sextupole and decapole corrections;
for main quadrupole: dodecapole corrections;

Other correction magnet are needed for requirements of beam dynamics, e.g.

Tune adjustment

Frequently, in machines with superconducting magnets, the main dipoles and quadrupoles are put in series for cryogenic economy of the current leads at operation currents of up to 5-15 kA in the transition from room temperature to cryogenic temperature, e.g. in HERA[1]; then, the tune of the machine is fixed; corrections needed are:

separate families of focusing and defocusing quadrupole correctors for tuning of the horizontal and vertical betatron oscillations.

Chromaticity correction

The focalization with the quadrupole magnets is obtained for a fixed value of the momentum of the beam particles; corrections needed are:

sextupole correctors to accomodate for the momentum spread of the beam particles; these should be located, where the dispersion is non-zero.

Various other correction elements are frequently used.

Additional elements

These may be needed for correction of other effects, e.g. coupling of horizontal and vertical betatron ocsillations, or injection and/or extraction of the beam, e.g.
skew quadrupoles, skew sextupoles, octupoles, etc.

Transition to straight sections and straight sections

Correction elements of similar types as above are also needed in the transition to the straight sections and in the straight sections themselves.

GENERAL REQUIREMENTS FOR CORRECTORS

Individual or grouped excitation

Correctors are excited individually or in families depending on their function. Typically, the horizontal and vertical dipoles in each cell used for orbit correction are excited individually. The quadrupoles for correction of magnetic length, tuning, saturation of the iron, and persistent current effects, and the sextupoles for correction of chromaticity, saturation of the iron, and persistent current effects are excited in families in each arc of the machine between straight sections.

Current requirements

Obviously, the heat load on the refrigeration system due to the current leads should be kept to a minimum. Also, the power dissipation of the bus bars at room temperature may be a non-negligible heat source in the machine tunnel. Therefore, the maximum excitation current for individually excited correctors is kept preferably below 50A, and for families of correctors below 100 A. Both polarities of the current should be available. This implies for superconducting correctors in most cases the design of coils with single strand cable.

Individual or grouped correctors

For saving of longitudinal space in the machine, groups of correctors of different multipole type are often made. This requires a detailed study of force patterns[2] and quench behaviour for all permutations of maximum excitation of both polarities. The forces may cause deviations from e.g. the cylindrical symmetry at zero excitation, which may also introduce new, unwanted multipoles[1].

Lumped or distributed correctors

Lumped correctors are separate elements which occupy longitudinal space in the machine lattice. Hence, they limit the longitudinal space available for main dipoles, and, thus, the maximum energy of the machine.

Correction dipoles are usually of the lumped type, as the Lorentz forces between this dipole and the main magnets would become very large.

Distributed correctors are mounted on the beam pipe inside the main dipoles and/or quadrupoles, and, hence, require transverse space which increases the overall transverse dimensions of the main magnet and, thus, the cost. Distributed correctors may be placed conveniently at the most effective position, e.g. sextupole correctors inside main dipoles for correction of persistent current effects.

Such correctors are limited by the forces between main magnet and corrector, mutual quench sensitivity, and mutual induction in case of a quench. E.g. the torque between a horizontal correction dipole and a main vertical dipole, or that between a skew correction quadrupole and a main normal quadrupole

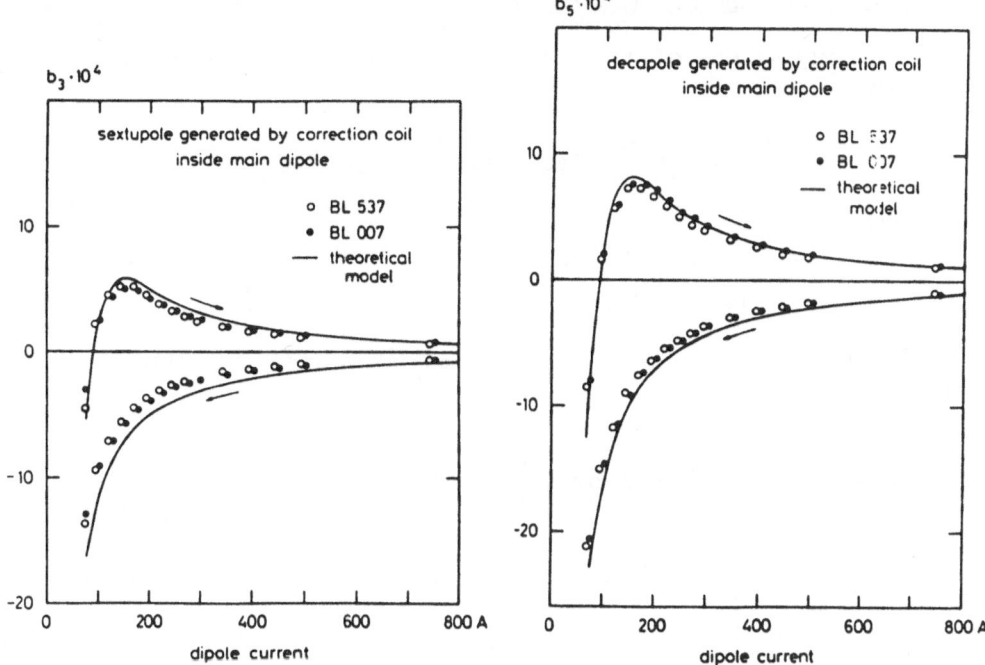

Fig.1.Sextupole and decapole generated by correction coil in main dipole[3]

is too large. A quench or more general a failure of a "trivial" corrector puts a costly main unit out of business.

Furthermore, the superconducting wires of the corrector are magnetized inside a main field due to persistent eddy currents induced inside the filaments. E.g. a sextupole correction coil inside a main dipole may have a decapole component of the same order of magnitude as that of the main field at the injection energy (HERA). This is shown[3] for the HERA dipole magnets in Fig.1. This can be reduced by using as fine filaments as possible in the conductor of the corrector.

The quadrupoles for correction of magnetic length, tuning, saturation of the iron, and persistent current effects, and the sextupoles for correction of chromaticity, saturation of the iron, and persistent current effects can either be of the lumped type or of the distributed type.

Grouping of correctors of same multipole type

As already indicated implicitly above, certain multipole correctors for different functions are grouped in one corrector, e.g.the quadrupoles for correction of magnetic length, tuning,saturation of the iron, and persistent current effects, and the sextupoles for correction of chromaticity, saturation of the iron, and persistent current effects.

Stored energy and selfinductance

The magnitude of the field of a 2p-pole with radius a is given by

$$B(r) = \begin{cases} B(a) \left(\frac{r}{a}\right)^{m-1} & \text{for } r<a, \\[2em] B(a) \left(\frac{a}{r}\right)^{m-1} & \text{for } r>a, \end{cases}$$

The stored magnetic energy of a magnet of length l is given by

$$U_m = \frac{1}{2\mu_0} \int (B(r))^2 \, 2\pi \, r \, dr = \frac{\pi l}{\mu_0} (B(a))^2 \left(\frac{a}{p}\right)^2 = \frac{1}{2} L \, i^2.$$

The stored energy is usually of the order of some tenths of J. For currents i of ~50 A, the selfinductance L is of the order of 10 mH.

DESIGN OF CORRECTORS

Correctors can have the following structures:

active correctors,

a) based on harmonic coils, like main dipole and quadrupole,
b) superferric magnets,

and passive correctors.

Active correctors

a) Harmonic coils.

These structures are subdivided in

single layer coils, like HERA 4-, 6-, 10-, and 12-pole correctors;

multilayer coils, like TEVATRON "spool" pieces.

Single layer coils are made out of thin current shells of superconductive single strand conductors (wires) mounted on the beam pipe parallel to the beam axis.

For multiwire single layer shells in a symmetric 2p-pole configuration at radius a, of thickness Δa, between azimuthal angles φ_1 and φ_2 with respect to the x-axis in the xy-plane, the multipole components in the horizontal plane (r=x, θ=0) are

$$B_{y,n}(x,0) = -\frac{\mu_0 \, j}{\pi} \, \frac{2p}{n} \left(\frac{x}{a}\right)^{n-1} \Delta a \, \left(\sin\varphi_2 - \sin\varphi_1\right) \delta_{n,(2k+1)p}.$$

Here, μ_0 is the permeability of the vacuum, p the basic multipole of the 2p-pole harmonic coils, j the current density, and $k = 0,1,2,..$.

For single wires in a symmetric 2p-pole configuration at angle φ, we have

$$B_{y,n}(x,0) = -\frac{\mu_0 \, j}{\pi} \frac{2p}{n} \left(\frac{x}{a}\right)^{n-1} \Delta a \, \cos(n\varphi) \, \delta_{n,\,(2k+1)p}.$$

Multilayer coils with inner radius a_1 and outer radius a_2 are designed as the main dipole and quadrupole magnets and the multipole components are

$$B_{y,n}(x,0) = -\frac{\mu_0 \, j}{\pi} \frac{2p}{n} (\sin\varphi_2 - \sin\varphi_1) \frac{x}{2-n} \left(\left(\frac{x}{a_2}\right)^{n-2} - \left(\frac{x}{a_1}\right)^{n-2}\right) \delta_{n,(2k+1)p} \, ,$$

for $n \neq 2$, and

$$B_{y,n}(x,0) = -\frac{\mu_0 \, j}{\pi} \frac{2p}{n} (\sin\varphi_2 - \sin\varphi_1) \times \ln\left(\frac{a_2}{a_1}\right) \delta_{n,(2k+1)p} \, ,$$

for $n = 2$.

The design of correctors has the same limitations as that of the main magnets, i.e. pure harmonic coils cannot be made. Correctors have also multipoles higher than the basic multipole type of the corrector. If we require, that the high multipoles of the correctors per lattice cell should be less than 2 units of the main dipole field integral per lattice cell, them the high multipoles of the field integral of the corrector itself with respect to its own basic field should be less than about 1%.

Superferric magnets

Superferric dipole correctors and some superferric quadrupole correctors are used in HERA. The field shape and quality is determined by the iron return yoke. These magnets have to be designed using programs like POISSON and TOSCA. Such magnets should be operated below the saturation of the iron to avoid unwanted high multipoles.

Passive correctors

These have been suggested by Brown and Fisk[4] for sextupole correction of persistent current effects in main dipole SSC magnets at low field. They put a pattern of strips of superconductor just inside the coil windings of the main dipole. The magnetization of these strips caused by the ramping of the magnet can compensate the persistent sextupole and decapole of the main dipole at injection by a proper choice of strip pattern and superconductor. This method has been demonstrated experimentally[5]. Requirements of dynamic range may limit the applicability of this method.

EXAMPLES OF CORRECTION MAGNETS

TEVATRON

The basic half cell of the Tevatron consists of four dipoles, a quadrupole, and a spool piece. The spool piece[6] is a group of lumped correctors with concentric multipole elements for saving longitudinal space. The following spool piece packages are used: DSQ, which contains a dipole, a sextupole, and a quadrupole with either horizontal or vertical dipoles; OSQ, which contains an octupole, a skew sextupole, and a skew quadrupole. Hence, in total three different packages are used.

The specifications of the various elements at $x_0 = 25.4$ mm are:

correction	strength	
dipole	0.47	Tm
quadrupole	0.17	Tm
sextupole	0.14	Tm
skew sextupole	0.11	Tm
octupole	0.08	Tm

The excitation currents for these values are of the order of 50 A. The expected excitation range is ±25 A, the design range is ±50 A, which for each coil is about 25% of the short sample limit at about 215 A. Fig.2 shows the cross section and the assembly of a DSQ package.

Design considerations and fabrication are as follows. The conductor has 0.02" diameter with a Cu to SC ratio of 1.45 to 1. This diameter is large enough for operation at about 25% of the critical current density. The insulation is Omegaclad, a multiple pass coating of polyesteramide/polyesterimide. (Of these, polyesterimide has the better radiation hardness). The coil consists of a large number of turns of small diameter wires, which are randomly wound. The mechanical constraints of the randomly wound coil are poor. Tight banding would cause a non-uniform distribution which can cause failure of the insulation or of the wire itself. Wire support is achieved by epoxy impregnation.

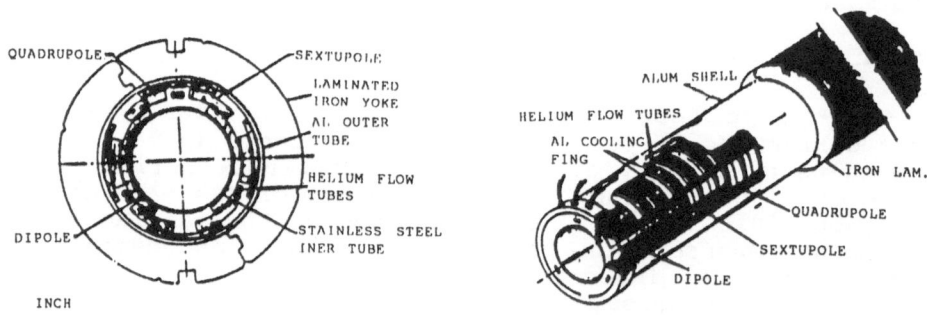

Fig.2. Cross section and assembly of a DSQ package[6]

For the impregnation various materials have been tested, polybondex and B-stage epoxy coating of the wire for selfbonding under heat treatment, and paraffin wax and Stycast 2850 FT, which is highly filled with Al-powder for high heat conductivity and crack resistance. The last was finally used.

Each coil element within a package will train in a manner typical of potted coils. After 5 to 10 quenches the coil will reach 80 to 85 % of the short sample limit which is about 215 A for each of the coils. When the coils are operated simultaneously, however, the operating current is sharply reduced.

HERA

The basic half cell of HERA consists of four dipoles, a quadrupole, and a correction dipole. For the transition to the straight sections quadrupole correctors are added. Correction coils of sextupole and quadrupole windings are put on the beam tube of the main dipoles[1].

The correction requirements at $x_0 = 25$ mm are:

correction	strength	
dipole	0.68	Tm
quadrupole (tuning)	0.47	Tm
quadrupole (arc to straight)	0.54	Tm
sextupole	0.35	Tm
decapole (persistent current)	0.005	Tm
(persistent current)	0.0005	Tm

The correction dipoles are superferric magnets with two saddle shaped coils[7]. The dipole has a length of 65 cm and an outer diameter of 25 cm, and is placed in the cryostat of the main quadrupole magnet. Fig.3 shows a side view and the cross section of the correction dipole. The coils are wound randomly on a flat race track mould. It has 1000 windings of 0.60 mm superconducting wire including a polyesterimide varnish insulation of 20 mm thickness. This varnish has a good radiation hardness[8]. The wire has a copper to NbTi ratio of 3.7:1 and contains 36 filaments of 45 μm diameter. During winding, the wire is wetted by epoxy (Epikote 215 and Versamid 140 in ratio 1:1 filled with 40 volume percent of Al_2O_3 powder). After winding, the long straight sections and the bends are constrained by compression bars. The centres of the short straight sections are clamped in fibre glass brackets. Then, the long straight sections are folded upward, and a natural bending of the unconstrained parts of the short straight sections at either side of the central bracket is obtained leading to a saddle shaped coil. The coil is baked out at 150^0 C for two hours. The coil is mounted in one half of a laminated iron yoke. Two half yokes are mounted and welded together to complete the dipole.

Fig.4 shows the distribution of quench currents before and after installation of the dipoles in the cryostat of the main quadrupoles. The maximum current in fig.4 before installation is essentially the short sample limit of the superconductor. After installation the maximum current is limited by the current carrying capacity of the current leads into the cryostat.

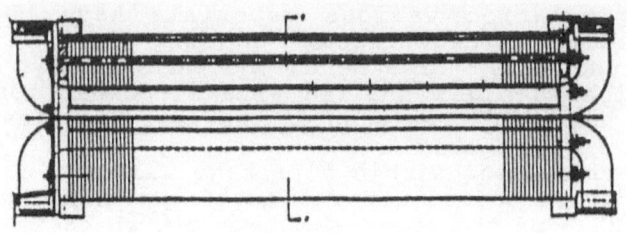

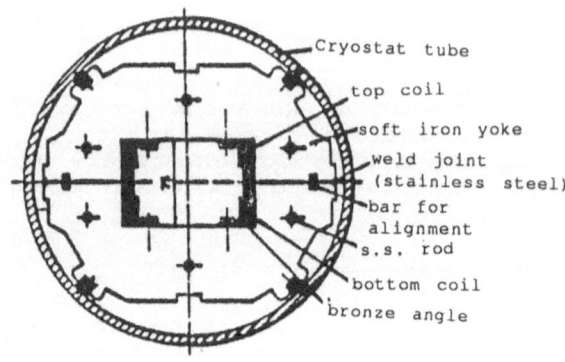

Cryostat tube
top coil
soft iron yoke
weld joint
(stainless steel)
bar for
alignment
s.s. rod
bottom coil
bronze angle

Fig.3.Side view and cross section of the HERA superferric correction dipole[7]

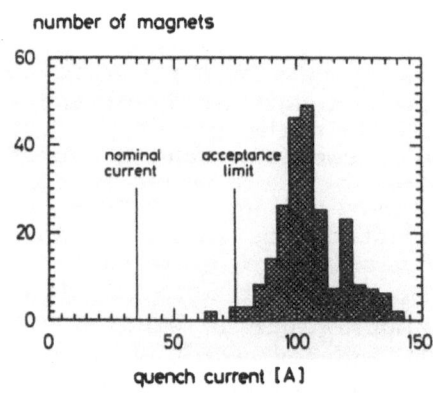

number of magnets

nominal
current

acceptance
limit

quench current [A]

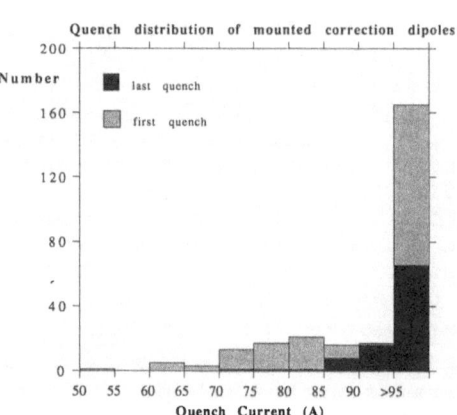

Quench distribution of mounted correction dipoles

Number

last quench

first quench

Quench Current (A)

Fig.4.Distribution of quench currents for the superferric correction dipoles before[7] and after installation in the cryostat of the main quadrupole

The superferric quadrupoles are produced in a similar way except that the coils are flat race tracks and have a regular winding pattern of rectangular superconducting wire of $0.45*0.90$ mm^2. The coil has 15 layers of 21 turns each. The wire has 24 filaments of 45 μm. The copper to NbTi ratio is 6:1. The insulation is polyesterimide varnish. Fig.5 shows the cross section of the superferric quadrupole, and the distribution of the quench current. The magnet has a length of 100 cm, and a diameter of 25 cm, and is placed in the cryostat of the main quadrupole magnet.

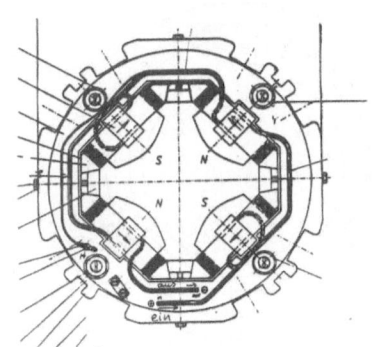
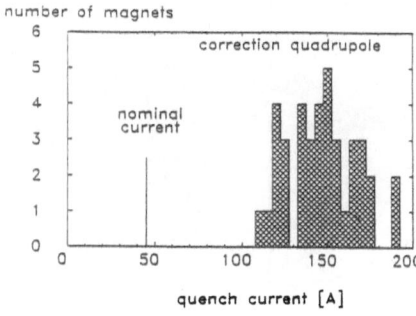

Fig.5.Cross section and distribution of the quench current of the HERA superferric correction quadrupoles

The design of the sextupole/quadrupole coils is based on that of the trim coils of Sampson and Cottingham for ISABELLE and CBA at BNL. The method of production has been modified. The coils are wound directly on a curved surface with the required diameter which is smaller than that for ISABELLE and CBA. The superconducting wire has a diameter of 0.7 mm and has an insulation of kapton and fibre glass with B-stage epoxy. The overall diameter is 0.83 mm. The wire has filaments of about 15 μm diameter. The copper to NbTi ratio is 1.8:1. The coil package consists of an inner sextupole layer of 5900 mm length and a outer quadrupole layer of 5830 mm length on one side of the 9582 mm long beam pipe of the main dipole magnet. The sextupole coil consists of three subcoils of 21 windings each, which are connected in series.

Each subcoil covers an azimuthal angle of 100^0 with a central G11 core of 20^0. A quench of the main dipole magnet induces a voltage in each subcoil, but no voltage over the string of three subcoils. The quadrupole coils consist of two subcoils of 33 windings each, which cover an azimuthal angle of 150^0 with a central G11 core of 30^0. With these angular ranges, the first harmonics are put to zero, and higher harmonics are sufficiently small. Again, a quench of the main dipole does not induce a voltage over the two subcoils in series.The coils are wetted with epoxy (Epikote 215 and Versamid 140 in a ratio of 1:1) and baked out at 150^0 C for two hours. Glass-kapton-glass tape layers of 0.15 mm thickness surround the beam pipe and are put in between the two coil layers for insulation. Each layer is glued with the same epoxy to the previous layer. The sextupole coils are surrounded by a single layer, the quadrupole coils by a double layer of very strong glass fibre (0.31 mm^2 VETROTEX R glass, tensile strength 3600 N/mm^2) with a tension of 800N/mm^2 and a pitch of 2.5 mm. This provides a high radial pressure of about 6 N/mm^2 at 4^0 K which together with the glue joints inhibit conductor motion under the influence of the Lorentz forces.

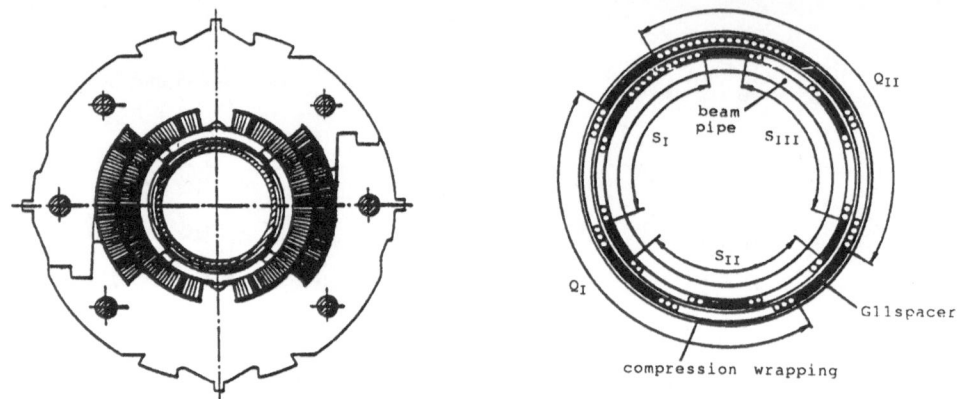

Fig.6.Cross section of the main dipole coil and the beam pipe with the sextupole/quadrupole correction coils, and the correction coil assembly:sextupole/quadrupole in inner/outer layer

Fig.6 shows a cross section of the main dipole with the correction coils and a cross section of the correction coils. Fig.7 shows an un-wrapped view of the three sextupole subcoils, and the two quadrupole subcoils.

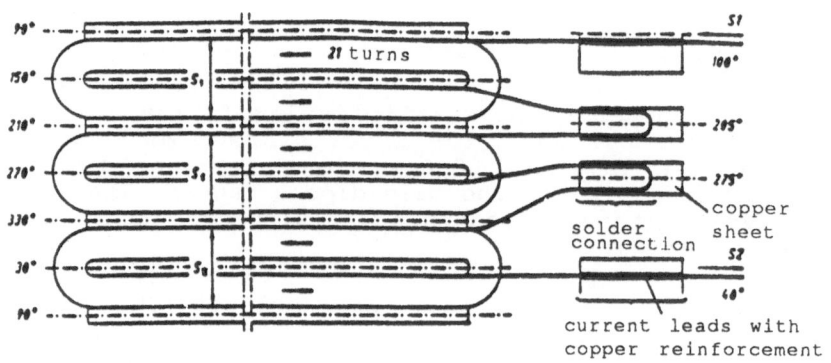

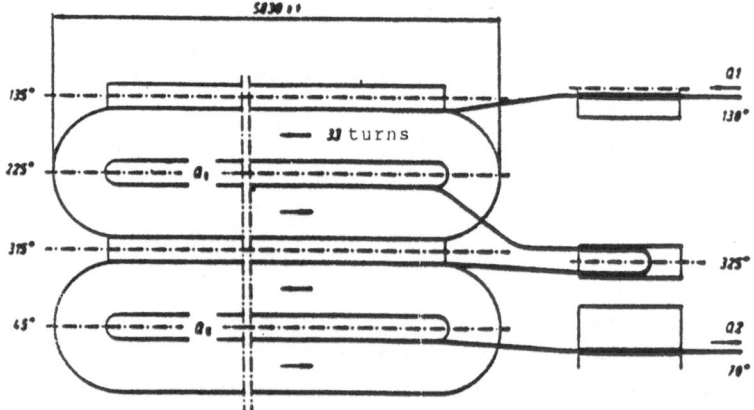

Fig.7.Unwrapped view of the sextupole, and the quadrupole coils, showing the subcoils

The relative orientation between the sextupole coil and quadrupole coil is measured in the warm field measurement of the multipoles at the end of the production. For all 460 sextupole quadrupole coils the deviation is (1.5 ± 1.5) mrad. For installation of the coils in the main dipole magnet, the sextupole subcoils SI ans SIII (see fig.7) have voltage taps. The main dipole magnet is excited with a low frequency source. The difference of the induced signals in these two subcoils provides a sensitive null method (~ 10 mV/mrad) for alignment.

Originally, TWARON, an aramide fibre, was used for the compression wrapping. It has a negative temperature expansion coefficient, and, therefore, the prestress is reduced after cooldown. Further, also plastic creep occurs already at room temperature. Vetrotex R glass does not show such behaviour, and was chosen for the compression wrapping. Fig.8 shows this for Kevlar 29, another aramide fibre with same properties as TWARON, and Vetrotex R glass.

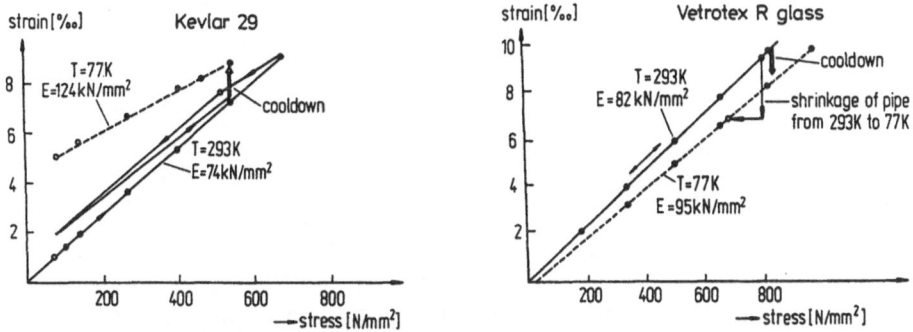

Fig.8.Stress-strain diagram of Kevlar 29 and Vetrotex R Glass at room and liquid nitrogen temperature

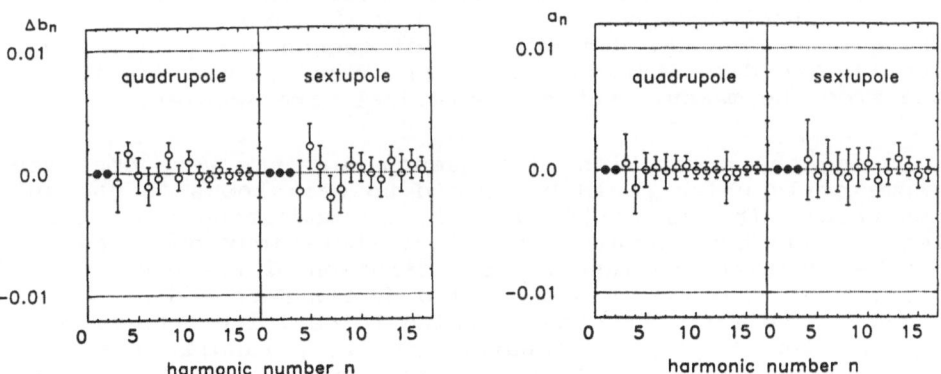

Fig.9.Normal and skew multipole coefficients of all 460 sextupole/quadrupole correction coils, for the normal coefficients the difference between measured and calculated values is plotted

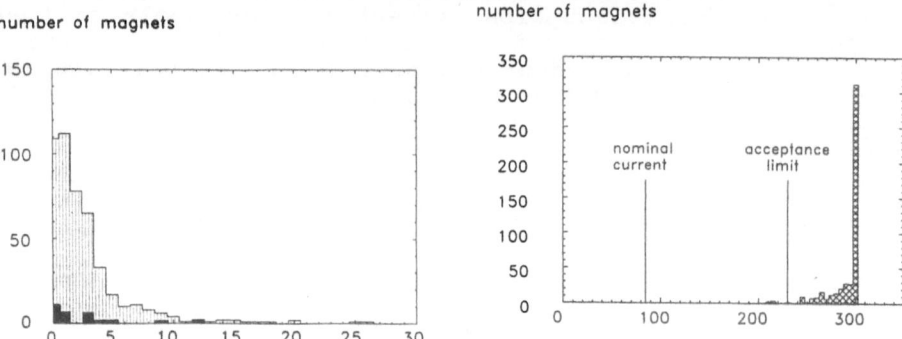

Fig.10.Number of training steps, and quench currents of all 460 sextupole/quadrupole correction coils.

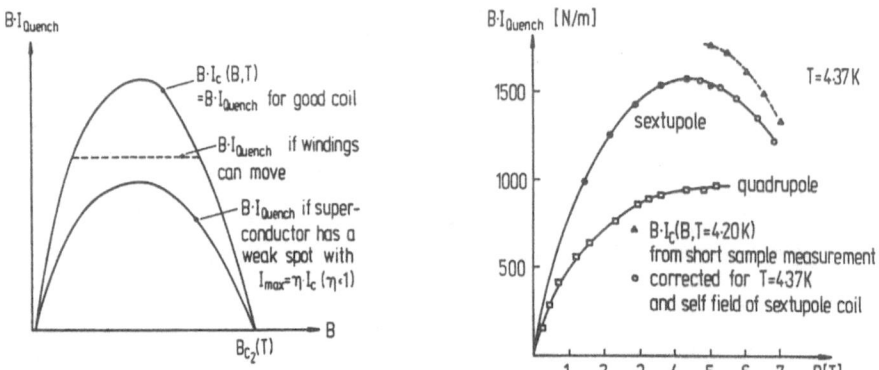

Fig.11.The product $B.I_C(B,T)$, resp. $B.I_{quench}$ for a) an undamaged superconductor, a coil limited by conductor motion, and a coil limited by a weak spot in the superconductor with $I_{max} = \eta I_c$ ($\eta < 1$), and b) the good/bad sextupole/quadrupole layer of coil B38.

Fig.9 shows the normal and skew multipole coefficients of all 460 sextupole/quadrupole coils. Fig.10 shows the number of training steps and the distribution of quench currents for all 460 coils. The distribution reaches the short sample limit, obtained from the measurement of individual wire samples.

During the production, low quench currents were observed for some coils which could be traced to weak spots in the superconductor. It was verified first by measuring the quench current versus temperature. The linear behaviour observed can hardly be explained by insufficient fixation of the windings. A second method, proposed and performed by ten Kate and Boschman, is based on the fact that the "pinning force", i.e. the product of I_{Quench} and B, shows a characteristic parabolic shape as function of the external field B. This product is the Lorentz force on the windings. Bad fixation of the windings should show a flat cutoff of the parabola. A weak spot in the superconductor shows a decrease of height of the parabola compared to the expected height from short sample measurements.This effect was observed. Fig.11 illustrates this behaviour.

Prototypes of the correction elements have been developed by Daum and Schmüser. The production has been performed by HOLEC in the Netherlands. However, the superferric quadrupole correctors were developed and produced at DESY by Schmüser.

To alleviate for persistent current effects at injection, it was necessary to correct also for the persistent decapole in the main dipole, and for the persistent dodecapole in the main quadrupole. Simple single decapole windings have been put on the part not occupied by the sextupole/quadrupole coils of the beam pipe of the main dipole over 2.9 m, and single dodecapole windings on the beam pipe of the main quadrupole over 2.0 m. These windings are put in between G-11 positioning strips on the beam pipe and fixed with a Vetrotex R glass wrapping as for the sextupole/quadrupole coils. The alignment of the decapole coils with respect to the sextupole coils on the same beam pipe is made with a magnetic measurement similar to that of the alignment of the sextupole coils in the main dipole. The dodecapole coils in the main quadrupole are aligned mechanically. In general, the magnetic measurement is more precise and reliable than the mechanical measurement. The decapole and dodecapole coils were developed by Schmüser and produced by Loll and Noell, respectively.

UNK

The basic half cell of UNK consists of four dipoles, and a quadrupole. Little information is available on the correction system of UNK. The main dipole magnets have cable with 6 μm filaments. The injection energy is 400 GeV and the final energy 3 TeV. For a top field of about 5 T, the field at injection is 0.67 T, which is relatively high. Then, persistent current effects are small, in particular for these thin filaments. Hence, sextupole and decapole correction for persistent current effects may not be needed.

SSC

The basic half cell of SSC[9] consists of five dipoles, a quadrupole, and a spool piece. Two types of spool pieces are foreseen. Primary correction packages are placed in each half cell adjacent to the main focusing (defocusing) quadrupoles with a horizontal (vertical) dipole for orbit correction, a focusing (defocusing) quadrupole for tune adjustment, and a focusing (defocusing) sextupole chromaticity correction. Twenty secondary packages are placed adjacent to the primary packages where needed. Detailed designs of the spool pieces have not yet been presented.

The suggested maximum strength of the correction elements at $x_0 = 10$ mm are:

Primary correction packages:	strength
dipole	3.10 Tm
quadrupole	0.28 Tm
sextupole	0.11 Tm

Secondary correction packages:

quadrupole	0.49 Tm
skew quadrupole	0.88 Tm
sextupole	0.50 Tm
skew sextupole	0.46 Tm
octupole	0.26 Tm
skew octupole	0.52 Tm

Originally, distributed sextupole correctors on the beam pipe of the main dipoles were planned for compensation of the persistent current sextupoles. Now preference is given to lumped correctors.

RHIC

The basic half cell of RHIC[10] consists of one dipole, a quadrupole, and an arc corrector. The arc correctors consist of four different multipoles: horizontal/vertical dipoles, normal/ skew quadrupoles, octupole, and decapole correctors. Separate sextupole elements correct for chromaticity and persistent current effects in the main dipoles.

The correction requirements at x_0 = 25 mm are:

correction	strength	
dipole	0.300	Tm
quadrupole	0.375	Tm
octupole	0.009	Tm
decapole	0.009	Tm
sextupole (chromaticity and persistent current)	0.188	Tm

The arc correctors are 0.56 m long packages. Excitation currents run from 160 to 200 A. The arc correctors are concentric around the beam pipe. From the centre out, the coil structures are decapole, octupole, and quadrupole, each with two layers of wires, and dipole with three double layers. The limiting angles are chosen to eliminate the first harmonic for the first three coils. For the dipole, the limiting angles of the three double layers are designed to minimize the higher harmonics. Each layer is wound on a flat substrate using the MULTIWIRE process[11]. This permits regular winding of the coils. The wire has 0.381 mm diameter, the Cu to superconductor ratio is 2.1:1, and has filaments of $\leq$ 9 μm diameter. The wire has a double wrapping of 25 mm Kapton. Each coil structure is secured to a stainless steel support tube. The sextupole corrector has to be strong, and, therefore, is a separate unit which is a superferric magnet. The coils consist of 200 turns of 0.5 mm NbTi multifilamentary wire. Fig.12 shows a cross section of the arc corrector and the sextupole corrector.

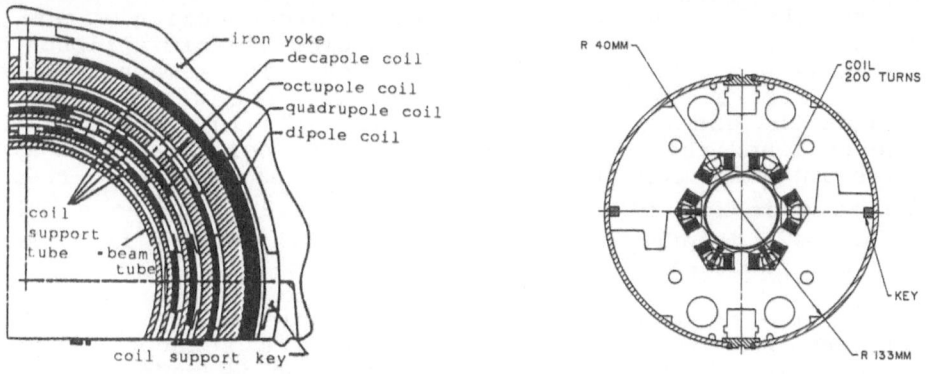

Fig.12.Cross section of the arc corrector and the sextupole corrector

LHC

The basic half cell of LHC[12] consists of four dipoles, a quadrupole, a tuning quadrupole, and a combined sextupole and dipole corrector. A separate sextupole corrector is foreseen for compensation of the persistent currents of the main dipoles.

The correction requirements at x_0 = 20 mm are:

correction	strength	
dipole	1.5	Tm
quadrupole	1.7	Tm
sextupole (chromaticity)	2.6	Tm
sextupole (persistent current)	0.042	Tm

The sextupole windings of the sextupole/dipole corrector are made of a rectangular superconductor of 0.7*1.2 mm^2. This permits regular winding of the coils. The nominal current is 458 A at 59 % of the short sample value. This is a high current for a correction element which seems to be the price which has to be paid for the choice of regular winding. Fig.13 shows a cross section of the dipole/sextupole corrector.

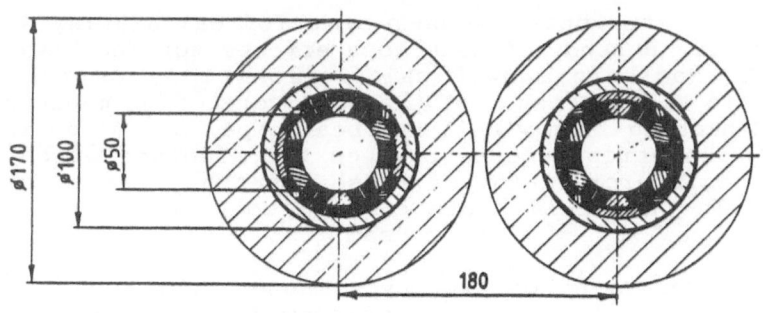

Fig.13.Cross section of the pair of dipole/sextupole correctors

The dipole windings have thin superconducting wires of 0.35 mm thickness. The wire is preassembled to form a ribbon of 12 parallel wires bonded together by epoxy. The ribbon is wound around a central hard copper island, and the wires are connected at the end to form a series coil. The inter-connections in the coil and the connections between the coils are made side by side in small copper tubes over 20 mm. The specific resistance is measured to be $13*10^{-8}$ Ωcm at 4.2 K giving a total heat loss (25 joints) of 3.6 mW. The nominal current is 47 A at 65% of the short sample value.

The superconducting wires of the dipole and sextupole are insulated with polyesterimide varnish. The sextupole/dipole corrector for chromaticity/orbit correction is under development by IJspeert[13] of CERN in collaboration with RAL, and TESLA.

A prototype of a pair of tuning quadrupoles is under development by Siegel[14] at CERN in a joint development project of the Spanish consortium ACICA and CERN(SPS/EMA). The design is based on that of the low-β quadrupoles of the ISR. It is intended to use a rectangular superconducting wire of about 3 mm^2 cross section for regular winding of the coils. Many details of the design are still being studied.

Sextupole and decapole correction coils for compensation of magnetization effects on the beam pipe of the main dipole have been designed by Cornuet[15] at CERN. It is another variant of the BNL trim coils. The superconducting wire is wound directly on the beam pipe in precise grooves of the extruded beam pipe. The grooves are lined with a triple glass fibre layer impregnated with B-stage epoxy. The wire has an insulation of polyesterimide varnish surrounded with glass fibre in the same B-stage epoxy. A compression winding of Vetrotex glass tape surrounds the package, which is subsequently insulated with kapton tape, and finally protected by a wrapping of Al wire.

However, in recent plans lumped correctors for magnetization effects are preferred for saving of transverse space, if superconductors with sufficiently thin filaments can be obtained.

EFFECTS OF BEAM LOSS

Several attempts have been made for calculating or devising tests for a study of beam loss effects for the Tevatron and for HERA. Hadronic shower calculations rely on complicated Monte Carlo simulations. For the HERA correction magnets, tests using DC heating from within the beam pipe, or pulsed injection of heat with strip heaters between the windings of correction coils. The interpretation of these tests have never been conclusive. In particular, distributed correction coils on the beam tube are exposed most to beam loss.

The development of prototype magnets for LHC should give a new chance to study effects of beam loss. It is intended to put a magnet or a string of magnets in an external SPS beam. These tests are very important for LHC, as it is intended to reach very high luminosities, $4*10^{34}$ cm^{-2}s^{-1}.

CONCLUSIONS

The general trend of the multiple choice for the correction elements seems to be the following. Orbit correction dipoles are usually executed as lumped elements. Tuning quadrupoles and chromaticity sextupoles are mainly taken to be lumped elements, except in HERA, where they are put as distributed elements on the beam pipe of the main dipole. Only in HERA, the chromaticity sextupoles compensate as well the persistent current sextupole of the main dipoles. Designs of new machines tend to use lumped sextupole correctors for the compensation of persistent current sextupoles, if conductore for the main dipole cables can be obtained with sufficiently small filament diameter. Elements are often grouped in one corrector.

The design of Eloisatron will profit from the experience obtained from the development and construction of the machines discussed in this contribution.

REFERENCES

1. C. Daum, J. Geerinck, P. Schmüser, R. Heller, H. Möller, M.Muto, S. Schollmeier, and P.A.M. Bracké, The Superconducting Quadrupolè and Sextupole Correction Coils for the HERA proton Ring, DESY HERA 89-09, Feb. 89.
2. C. Daum, and P. Schmüser, Quadrupole and Sextupole correction Coils for HERA, DESY HERA 83-01, March 1983.
3. P. Schmüser, private communication.
4. B.C. Brown, and H.E. Fisk, A Technique to minimize Persistent Current Multipoles in Superconducting Accelerator magnets, Proc. of the 1984 Summer Study on the Design and Utilization of the Superconducting Super Collider, Ed. R. Donaldson, and J.G. Morfin (1985) p.336.
5. H.E. Fisk, R.A. Hanft, M. Kuchnir, and A.D. McInturff, Passive Correction of Persistent Current Multipoles in Superconducting Accelerator Dipoles, Proc. of Workshop on Superconducting Magnets and Cryogenics, Brookhaven National Laboratory, May 12-16, 1989, BNL 52006, Ed. P.F. Dahl, p.222.
6. D. Ciazynski, and P. Mantsch, Correction Magnet Package for the Energy Saver, IEEE Transactions on Nuclear Science (1981), Vol. NS-28, No. 3, P. 3325.
7. C. Daum, J. Geerinck, R. Heller, M. Muto, S. Schollmeier, P. Schmüser, and P.A.M. Bracké, Superconducting Correction Magnets for the HERA proton Ring, in DESY HERA 88-09, June 1988, p. 1, Reports at the EPAC – European Particle Accelerator Conference, Rome, June 7-11, 1988.
8. P. Beynel, P. Maier, and H, Schönbacher, Compilation of Radiation Damage Test Data, Part III, Materials used around high-energy accelerators, CERN 82-10, 4 November 1982, p.144.
9. Conceptual Design of the Superconducting Super Collider, SSC Central Design Group, SSC-SR-2020, March 1986.
10. Conceptual Design of the Relativistic Hadron Collider RHIC, BNL52195, March 1989.
11. J. Skaritka, W. Schneider, R. Shutt, P. Thompson, P. Wanderer, E. Willen, D. Bintinger, R.Coluccio, and L. Schieber, Development of the SSC Trim Coil Beam Tube Assembly, 1987 Particle Accelerator Conference, Washington, D.C., 1987, p. 1437.

12. The Large Hadron Collider in the LEP tunnel. Eds. G. Brianti,and K. Hübner, CERN 87-05, 27 May 1987.
13. A. IJspeert, private communication.
14. N. Siegel, private communication.
15. D. Cornuet, private communication.

DYNAMIC APERTURE CONSIDERATIONS FOR LARGE SUPERCONDUCTING

SYNCHROTRONS

F. Willeke

Deutsches Elektronen-Synchrotron DESY

Notkestr. 85, 2000 Hamburg 52, Germany

INTRODUCTION

Nonlinear forces introduce a limit for stable betatron oscillation amplitudes of the particles of an accelerator beam. Beyond this limit, the particle amplitudes will grow until the wall of the beam pipe is reached and loss occurs. That is why the focussing and guide fields of accelerators are designed to be as linear as possible. On the other hand, a perfectly linear machine would not work. Nonlinear sextupole fields have to be installed in order to compensate chromatic effects[1] . This turns the linear lenses effectively into achromats. The impact of the nonlinearity of these sextupoles and corresponding performance limitations have to be tolerated. Thus in every large accelerator there is a limit for the amplitude of stable betatron oscillations. Chromaticity correcting sextupoles present a natural scale to determine limitations for additional nonlinearities which arise from imperfections of the magnetic guide and focussing fields. Unfortunately, the strength of these parasitic nonlinearities becomes comparable to or even larger than the chromaticity sextupoles if the synchrotrons become considerable larger than existing machines.

In this lecture, the basic concepts and procedures to arrive at appropriate and feasible specifications of tolerances on magnet field errors will be presented. The outline is as follows: In the first section the different aperture concepts which are used by accelerator designers will be presented. The second section reviews the impact of the different types of magnet field errors on the beam performance. Finally, since the high field quality requirements are in practice difficult to

---

[1] Chromatic effects are caused by the change in the focussing fields due to a small momentum error of the particles. The most important effect is the chromaticity. the change of the tune with momentum $\xi = \Delta q / \frac{\Delta p}{p}$.

meet by magnet builders, additional cures are needed to reduce these requirements. Correction elements and sorting schemes will be discussed in the last section.

APERTURE CONCEPTS FOR LARGE HADRON ACCELERATORS

The different concepts of apertures in the context of beam stability can be summarized in the following way:

Needed Aperture. A certain area around the central trajectory of the beam in a synchrotron is occupied by the particles which perform betatron oscillations around a central trajectory which is called the closed orbit. The size of the oscillation amplitudes depends on the emittance of the beam extracted from the ion source at the start of the acceleration chain and, in addition, on how precise one is able to steer the beam on the transfer from one accelerator to the next in this chain. Since there is a momentum spread in the beam and since particles with higher momenta travel on a somewhat larger radius and particles with smaller momentum on a somewhat smaller radius a somewhat larger aperture is filled. The closed orbit is not necessarily the design orbit which is centered in the beam pipe. There are closed orbit distortions which arise from misplaced quadrupole magnets or variation of the magnetic guide field. Since one needs first some beam in the machine to correct for these errors, some additional space is needed. Furthermore, in order to study the beam behaviour and thus to optimize the machine performance, the beam is sometimes kicked so that it oscillates as a whole around the central trajectory, or the whole beam is forced to travel somewhat on the inside or outside the center of the beam pipe. Furthermore, only in the ideal case, injection of the beam right on the central trajectory is possible. In practice, there are always residual oscillations which result from nonperfect adjustment of injection elements. Moreover, in order to be able to adjust injection parameters, we must admit injection of the beam under nonoptimized conditions and we must provide some additional free aperture to accomodate beams with poor injection quality. These considerations determine the needed aperture (see fig 1). It is given by the formula

$$A_{needed} = 2 \times \left(\sqrt{6 \left[\varepsilon\beta/\gamma + \left(D\frac{\Delta p}{p} \right)^2 \right]} + A_{op} \right) \times \text{Safety factor}$$

Here ε is the normalized beam emittance which encloses 65°% of the particles, ß is the focussing function, γ is the beam energy in unit of the rest energy ($v = c$ is assumed), D is the dispersion function, $\Delta p/p$ is the half width of the momentum distribution and A_{op} is the additional operational (half) aperture need. For a machine like the HERA proton ring ($\varepsilon = 4.5\mu m$, $\hat{\beta} = 80m$, $D = 2m$ and $\Delta p/p = S \cdot 10^{-4}$) one estimates a beam radius of $\sigma = 10mm$ (enclosing 95% of the particles). For operations one estimates an additional aperture of 5mm. This results in an aperture need of 30mm including a safety factor of $\sqrt{2}$ [1].

Dynamic Aperture. The dynamic aperture is the aperture inside which particles perform stable betatron oscillations. Thus a particle injected inside that aperture would circulate for ever around the accelerator. The dynamic aperture is the innermost border of a projection of a complicated shaped surface in 6-dimensional phase space on the transverse coordinate axis inside the beam pipe. The mechanism which causes an amplitude limit for stable motion can be very complicated. Well understood are resonant effects. The nonlinear force then oscillates with the betatron frequency which leads to a rapid growth of oscillation amplitudes. This resonant instability requires a minimum strength of the resonant component of the nonlinear force and certain resonant betatron tunes which satisfy

$$nQ_x + mQ_y + kQ_s \simeq integer.$$

($Q_{x,y,s}$ are the horizontal, vertical and longitudinal tunes, n,m,k are integers.) Resonances can be avoided by appropriate choice of the tunes or by a compensation of the resonant components of the nonlinear force. The dynamic aperture is then determined by more subtle effects. The trajectories of the particles beyond the dynamic aperture exhibit often a strange, irregular and unpredictable behaviour which is called "chaotic". The dependence of the oscillation frequency on the amplitude (called detuning) which is characteristic for nonlinear systems is one of the reasons for this behaviour. Since the tune changes with amplitude, different resonances are no longer separated but interfere with each other. This is called resonance overlap. It is a criterion for chaotic motion[2]. If the motion is chaotic, amplitudes may vary only little for quite a long time before a rapid growth sets in. It can take many turns (10^5 – 10^6 turns which corresponds to 2.1s—21s real time in HERA) before a particle loss occurs. It requires a large effort to predict the dynamic aperture by a computer model of the accelerator. Long time tracking calculations are a straight forward approach. An alternative method to determine whether a particle is stable or unstable is to detect the chaotic character of the motion. This reduces the numerical effort to determine the dynamic aperture. A criterion for chaotic motion is a nonvanishing Lyapunov exponent[3] which measures the density of unstable regions in phase space around the trajectory considered. The detection of an exponential divergence of two trajectories which are injected very close to each other in phase space is a test for a nonvanishing Lyapunov exponent. A pragmatic procedure to detect a nonvanishing Lyapunov coefficient has been developped and successfully used to detect the single particle stability in the HERA proton ring[4,5]. This method saves up to a factor of ten in computing time compared to long time tracking.

Tracking Aperture. Investigation of the impact of nonlinearities on the beam dynamics by computer simulation involves a scan of important parameters. Key parameters are the particle tunes, momentum amplitude, ratio of transverse amplitudes, the chromaticity, the strength of the nonlinearities, closed orbit errors and seeds of "random" distributions. They will be varied systematically and the dynamic aperture has to be determined for each parameter value. Though it is not hard to define dynamic aperture, it is quite difficult and time consuming to determine the dynamic aperture. Therefore an artificial dynamic aperture definition is used for systematic computer simulations. This is the tracking aperture. It is defined that inside the tracking aperture, the particles in the beam perform stable betatron oscillations for a restricted number of turns around the machine (100—1000). This criterion can be tested on a computer by relatively moderate means. (To simulate the motion once around HERA using the tracking code RACETRACK[25] with multipole errors taken into account in each of the 400 dipoles takes ~ 100ms per particle on an IBM 3090 computer). The tracking aperture turns out to be considerably larger than the dynamic aperture. They differ by a factor of approximately two according to tracking studies for the HERA proton ring[4]. The tracking aperture should therefore only be used for comparisons. One should not draw conclusions from its absolute value.

Life Time Aperture. Particles inside the dynamic aperture would circulate in the accelerator for an infinite time, if there was no external excitation as for example induced by power supply ripple, or interaction with the residual gas in the beam pipe. In reality, these effects are always present. They interfere with the nonlinearities. For example, tune modulation caused by power supply ripple introduces sidebands in the vicinity of the nonlinear resonances. Since there is detuning, these sidebands tend to overlap which broadens the chaotic regions in phase space (see for example[7]). Furthermore the complicated topology of nonlinear phase space leads to an enhancement of the excitation of betatron oscillation due to external noise (fast ripple, multiple scattering etc). Therefore, the prediction the dynamic aperture and stable motion only guarantees beam lifetimes in the order of seconds of real time. This is the result of nonlinear dynamic experiments and its comparison with simulations of the beam behaviour which has been performed in the SPS at CERN[8]. In that particular experiment, the motion had to be restricted to a considerably smaller aperture than the dynamic aperture in order to achieve life times in the order of hours (which is required in a storage ring). This aperture may be called the life time aperture. Since it requires an extremely large amount of computing time to simulate the motion of a particle for macroscopic times, and since there is a wide range of possible effects which are relevant for the beam stability, the life time aperture and its relationship with tracking aperture and dynamic aperture can only be determined by machine experiments. Thus experiments on existing machines play an important role in the design of a new machine.

Linear Aperture. A quite different approach to arrive at tolerances for magnet field errors is the concept of linear aperture. The basic idea is that the needed aperture should be inside a region where the beam behaves sufficiently linearly, thus where nonlinear field errors cause only small tolerable distortions of the motion as described by the linearized equation of motion. The motion of the beam can then be predicted well enough by a linear analytic model (linear beam optics). Most of the procedures to control the beam and to optimize its performance will only work properly if the motion obeys the linear optics. (One has to think about orbit corrections, chromatic compensations, tune adjustments, feed back systems etc.) Moreover, if the nonlinear fields are sufficiently small, the distortions can be described with sufficient accuracy by perturbation theory and the nonlinear impact on the beam dynamic becomes transparent. This enables us to install and adjust appropriate correction elements. Such an almost linear motion should also be stable. However it is not so obvious what sufficiently linear means. In order to test linearity it has been proposed[9] to use the following two criteria, the rms deviation of betatron oscillation amplitudes from constant linear amplitudes (this has been called "smear") and the amplitude dependence of the tunes. The motion in the Superconducting Super Collider (SSC) is considered sufficiently linear if the variation of betatron oscillation amplitudes is smaller than

$$S_y = \frac{\sqrt{\langle a_y^2 \rangle - \langle a_y \rangle^2}}{\langle a_y \rangle} \leq 10\% \tag{1}$$

(where a_y^2 is the Courant Snyder invariant $a_y^2 = \gamma y^2 + 2\alpha yy' + \beta y'^2$, y denotes either horizontal or vertical motion) and the tunes differ from the linear tunes only by

$$\Delta Q_y \leq 0.005 \tag{2}$$

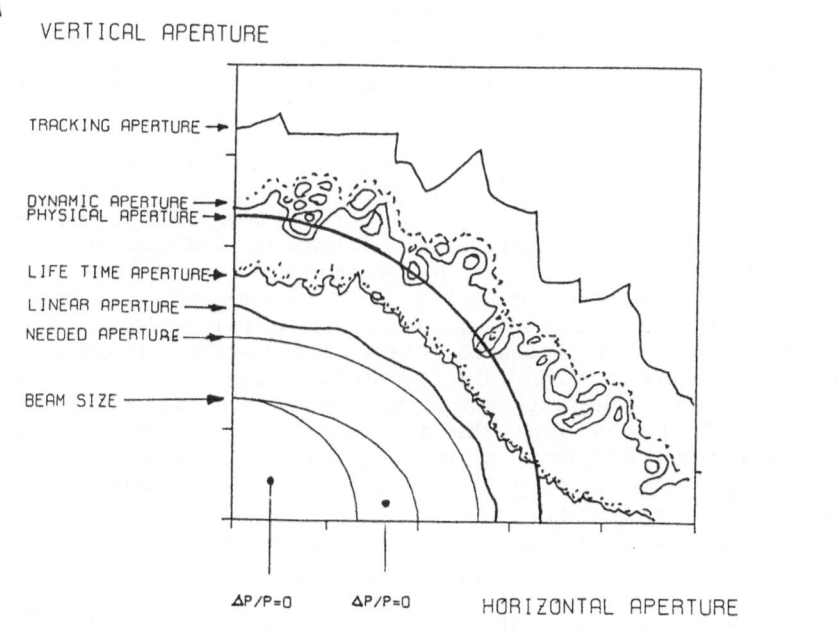

Figure 1. Schematic Comparison of Various Apertures

Experimental tests in the SPS at CERN however show that the linearity criterion for the SSC does not guarantee that the beam is inside the dynamic aperture and does not guarantee that the motion is stable[8]. The conclusion is that simple global criteria such as the "smear" and the tune shift with amplitude have to be very restrictive in order to be a safe criterion which assures the beam stability. The reason is that two situations which are qualitatively completely different may be characterized by the same value of tuneshift and smear. It is therefore not recommended to use smear and tuneshift as global stability criteria.

MAGNETIC FIELD ERRORS AND SINGLE PARTICLE STABILITY

One distinguishes between systematic and nonsystematic field errors.

Systematic "Geometric" Field Errors

Field errors which are conformal with the symmetry of a dipole magnet are mirror symmetric with respect to the vertical coordinate axis in the center of the magnet. Multipole components which have this symmetry are called allowed components. In large regular structures which comprise the main part of a large superconducting synchrotron, the impact of allowed systematic nonlinear errors of the bending field cancels to a large extent intrinsically. The reason is that the distortion of the motion of a particle by the nonlinear force is partly cancelled by another force which acts farther downstream, where the particles have advanced by 180° in betatron oscillation phase. The intrinsic cancellation is not quite perfect because the betatron phase advance is slightly distorted by the nonlinear forces which act between the two points considered. Therefore, large systematic field errors nevertheless do have an impact. Moreover, only particles with the ideal momentum travel effectively in the center of the magnet. Particles with a momentum offset oscillate around an orbit which is shifted radially outwards or inwards, and they experience also antisymmetric forces. The impact of antisymmetric fields does not intrinsically cancel. There is always an average component which builds up coherently even in regular structures. This causes a strong amplitude dependence of the tune which is closely connected with the occurrence of chaotic motion which limits the dynamic aperture. As a consequence, the dynamic aperture tends to decrease monotonically with the momentum deviation of a particle (if a constant momentum deviation is considered). Fig.2 shows as an example the acceptance (square of the tracking aperture divided by the ß-function) for the TEVATRON plotted versus constant momentum deviation which shows this characteristic behaviour. The systematic field errors which are caused by nonperfect coil geometry, can be reduced considerably by integrating spacers in the superconducting coils as has been demonstrated successfully by the HERA superconducting dipole[10]. The spurious systematic high excitation field errors in the HERA dipole magnet as shown in table II[11] cause only an insignificant reduction the HERA dynamic aperture. The FERMILAB magnet however has a high 14-pole and 18-pole content in the dipole field (see table I)[12].

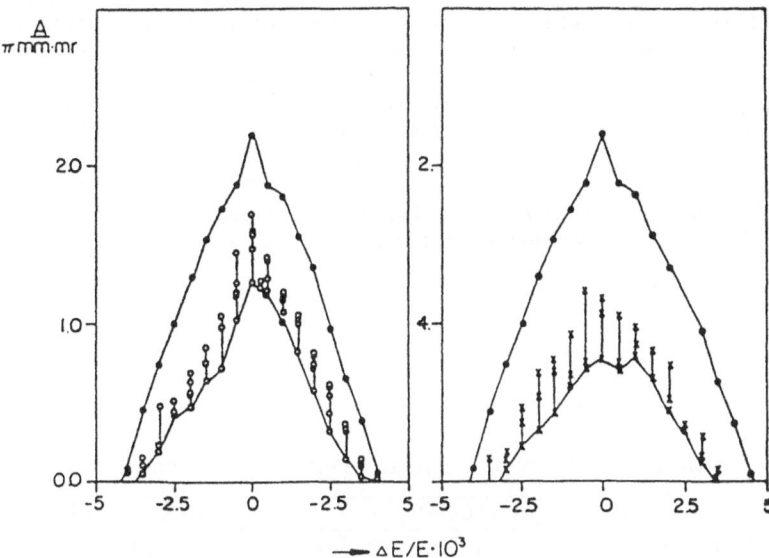

Figure 2. TEVATRON Dynamic Aperture Determined by Particle Tracking as a Function of Constant $\Delta p/p$ with Multipole errors in the dipoles,
left figure: o systematic errors only, o including random field errors,
right figure: + including orbit distortions $x_{rms} = y_{rms} = 2mm$ in addition

These field errors contribute considerably to the dynamic aperture limitation in the TEVATRON as has been shown by systematic tracking calculations[14,13] and by an analytic study of the nonlinear dynamics of the TEVATRON[15]. This underlines the importance of optimum coil design for the performance of a superconducting accelerator.

Nonsystematic Field Errors. The intrinsic compensation of the nonlinear impact does not occur for nonsystematic nonlinearities. There is some cancellation due to the randomly changing sign of nonsystematic field errors. This prevents a coherent build up of detuning terms which cause the amplitude dependence of the tunes. Nonlinear detuning effects excited by nonsystematic field errors are therefore small compared to the effect of systematic errors. (The reduction is in the order of $1/\sqrt{N}$, N being the number of distorted magnets). On the other hand, the potentially resonant harmonics of nonsystematic field errors are expected to be much larger than for systematic errors of the same magnitude. For these reasons, the nonsystematic field errors are harmful for the dynamic aperture. This is confirmed by many particle tracking studies for various machines. See for example the tracking results for the TEVATRON (fig.2) which show a considerable decrease of dynamic aperture if random fluctuations are superimposed on the systematic multipoles according to table I. Since the systematic field errors in HERA at high magnet excitation are very small, the high field dynamic aperture of $16mm^4$ in HERA is governed by the effect of the nonsystematic field errors.

Table I. Summary of Fermilab Magnet Measurements $\Delta B/B$

Low Field (Injection Energy 150GeV)				High Field (1TeV)				
Normal Component		Skew Component		Normal Component		Skew Component		
n	average	sigma	average	sigm	n	average	sigma	average	sigm
1	0.0	0.0	0.0	0	1	0.0	0.0	0.0	0
2	0.0	0.63	0.0	0.79	2	0.0	0.63	0.0	0.69
3	-4.71	3.75	-0.11	1.29	3	-0.99	3.50	0.38	1.29
4	-0.23	0.89	-0.04	1.67	4	-0.27	0.87	-0.07	1.69
5	0.12	1.42	-0.04	0.56	5	-0.76	1.60	-0.07	0.53
6	-0.02	0.41	-0.15	0.69	6	-0.05	0.37	-0.10	0.67
7	6.69	1.03	0.15	0.41	7	5.48	1.01	-0.13	0.46
8	0.02	0.29	0.25	0.44	8	-0.12	0.29	0.45	0.35
9	-15.69	1.98	-0.73	0.87	9	-15.43	2.01	-0.68	0.88
Values in Units of 10^{-4} at r=25.4mm								

Table II. Status of HERA Magnet Measurements $\Delta B/B$

Low Field (Injection Energy 40GeV)				High Field (820GeV)				
Normal Component		Skew Component		Normal Component		Skew Component		
n	average	sigma	average	sigm	n	average	sigma	average	sigm
1	–	–	–	–	1	–	7.00	–	–
2	0.01	0.69	-0.47	2.97	2	-0.05	0.67	-0.42	1.68
3	-32.28	3.08	-0.13	0.78	3	0.34	2.60	-0.28	0.45
4	0.14	0.43	0.70	1.38	4	0.16	0.27	0.21	0.90
5	11.72	1.36	-0.00	0.73	5	0.84	0.83	0.00	0.24
6	-0.12	0.38	-0.68	0.62	6	-0.06	0.11	-0.07	0.24
7	-1.98	0.65	-0.10	0.52	7	0.26	0.27	-0.01	0.12
8	0.11	0.41	0.20	0.41	8	0.02	0.07	-0.01	0.12
9	0.359	0.51	-0.01	0.57	9	-0.33	0.12	0.04	0.10
Values in Units of 10^{-4} at r=25mm								

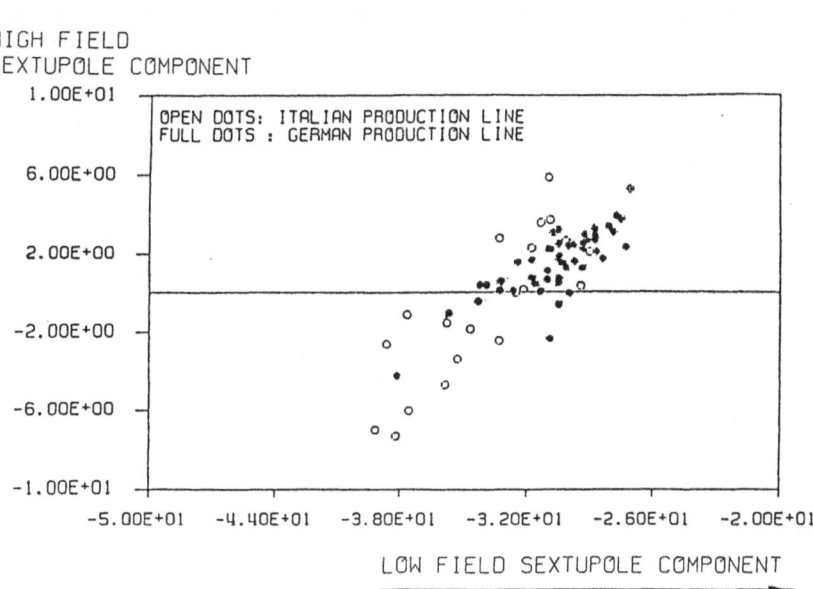

Figure 3. Correlation of High Field (I = 5000A) Sextupole Components and Low Field (I = 250) Sextupole Components in the HERA Superconducting Dipoles

Persistent Current Field Errors. Persistent currents in superconducting magnets drive very strong systematic magnetic field imperfections at low magnet excitation. The injection energy in the TEVATRON, the first superconducting large synchrotron, has been chosen sufficiently high, so that the persistent current field errors are not larger than the systematic geometric errors. More typical for future large machines is the HERA proton ring where the magnetic field has to change by a factor of 20 between injection and top energy. Thus one has to inject into a field where the magnitude of the persistent current field distortions is close to its maximum. Systematic magnetic field measurements for the TEVATRON and the HERA superconducting magnets confirm that there is no strong nonsystematic component of the persistent current field errors as expected. This becomes obvious if one compares low excitation field errors with high excitation field errors, which are summarized in tables I and II. See also fig.3, where the sextupole component of the HERA dipole field at high excitation (I = 5000A) is plotted versus the low excitation (I = 250A) sextupole field errors for all HERA dipoles measured so far.

The impact of persistent current driven field errors on the dynamic aperture is the same as for any other systematic dipole error. The field errors obey the constraints given by the symmetry of the coils. The extraordinary strength of these field distortions leads to a situation which is qualitatively different from the case with only geometric field errors.

In the HERA proton ring, the impact of the persistent current sextupoles which amounts to 0.3% of the dipole field (measured at r = 25mm) would completely destroy the dynamic aperture at injection energy in comparison to the value obtained for chromaticity correcting sextupoles. The strength of the sextupole requires a distributed correction (see below).

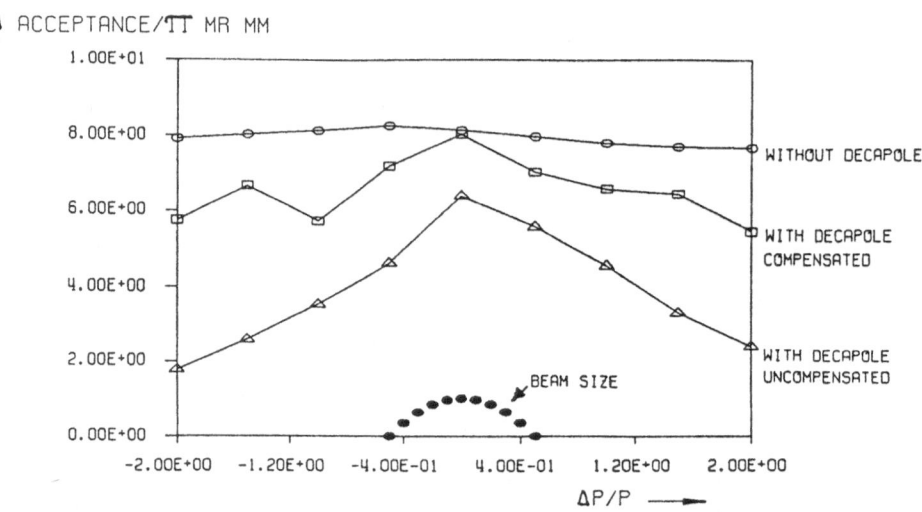

Figure 4. Nonlinear Acceptance of the HERA Proton Ring nonsystematic multipoles only(dots), with an additional decapole component of b_5 =−0.1% @ r = 25mm in the dipoles without compensation (triangles) and with lumped correctors (squares). The dotted line encloses the area occupied by the beam at injection

These corrections do not compensate the higher order multipoles induced by persistent currents. The decapole component of 0.1% at r = 25mm in the dipole magnet as well as the duodecapole component of 0.4% in the superconducting quadrupole magnets is still strong enough to cause an intolerable decrease of the dynamic aperture which is shown in fig.4 (which is taken from ref[16]). The ratio between persistent current sextupole and persistent current decapole only depends on the coil geometry. This is confirmed by comparison of FERMILAB and HERA magnet data (see table I and II) as well as by modeling the persistent field errors of superconducting magnets by computer calculations[17].

The persistent currents in the quadrupole magnets drive a strong duodecapole component. In the HERA superconducting quadrupoles it amounts to 0.4% of the quadrupole field at r = 25mm. The corresponding antisymmetric nonlinear field causes a tuneshift of $\Delta Q = 0.19$ for particles with oscillation amplitudes of 20mm (which corresponds to $\simeq$ 2/3 of the physical aperture). From these distortions results a reduction of the tracking aperture in the HERA proton ring by a factor of three[16].

Serious problems are also caused by the decay of persistent currents due to flux creep[18]. If the magnetic field in the superconducting magnets is held constant at low excitation, as will be necessary during injections the persistent currents which have been induced up to this point will decay slowly. The decay is well described by a logarithmic function of time. In the HERA dipole magnet, low excitation persistent current sextupole components decay by 3×10^{-4} @r = 25mm in about 30 minutes. What happens if one starts ramping again, is that the decayed persistent currents are reinduced very quickly. This has been observed in the TEVATRON[19] as well as in HERA magnet measurements (see for example[11]). The corresponding rapid change of sextupole strength would change the chromaticity in HERA by about 50% of its natural (uncompensated) value if one is not able to follow instantaneously by appropriate changes in the correction system. Under such conditions, it is very difficult to store protons in the machine. The TEVATRON, where the chromaticity change is only half the value which one expects for the HERA proton ring, suffered from bad beam lifetime and poor performance until this effect was recognized and cured. It is well known that tune modulations due to momentum oscillation with uncompensated chromaticity leads to a reduction of dynamic aperture by the mechanism described in the previous section. Tracking calculations have been performed for HERA for the case where the chromaticities from the rapid persistent current change are not compensated. A drastic reduction of the dynamic aperture has been predicted under these circumstances[20].

CURES

The field quality requirements which arise from beam stability considerations in superconducting synchrotrons cannot be met due to the persistent current problem. Moreover, if the synchrotrons get very large, other sources of magnet field errors exceed that which can be tolerated by the beam. Consider for example the tuneshift with amplitude excited by a systematic ("forbidden") octupole. For an amplitude which corresponds to the needed aperture A, and for an octupole strength of b_3 which is measured at the physical aperture r_0, for a lattice with a half FODO cell length L and a betatron phase advance of 60° per FODO cell one obtains

$$\Delta Q \quad \sim \quad 0.58 \cdot b_3 \left(\frac{A}{2r_0}\right)^2 \cdot \left(\frac{L}{r_0}\right) \tag{3}$$

Inserting the parameters for a large machine such as the ELOISATRON[21] (r_0 = 15mm, A/2 = 7.5mm = $r_0/2$, ,L = 200m) one arrives at a large tune shift of ΔQ = 0.019 if one assumes an octupole corresponding to the average octupole component which has been measured in the HERA dipole magnet $b_3 = 10^{-5}$@r = 25mm. Even such a small residual field imperfection causes a tune-shift which is four times larger as the one considered tolerable for the SSC. These numbers suggest that future large accelerators will not work without appropriate correction schemes.

Compensation of Systematic Field Errors. The most important systematic field errors come from persistent current sextupoles. One may avoid the persistent current problem field errors by raising the injection energy. This would require a much more powerful injector which might be the most expensive way of solving the problem. Installation of correction elements is another possibility. Obviously, the best correction system which can be installed is to put active correctors inside the superconducting magnets as has been done in case of the HERA dipoles. This however requires a larger magnet aperture in order to to provide the additional space which also means considerably higher costs of the magnets. Much effort has been spent in investigating correction schemes which use a few lumped correctors placed at appropriate points in the lattice. One of the schemes discussed is the so called Neuffer scheme[22]. The underlying beam dynamics considerations are representative for the many other correction schemes which are being discussed for the SSC[23]. In order to demonstrate how a lumped corrector scheme works, is is useful to consider a Fourier decomposition of the vector potential from which the nonlinear forces are derived. The Fourier components are obtained by integrating the nonlinear field coefficients b_k which are multiplied by appropriate powers of the betafunctions and a function with multiples of the betatron phase[2]

$$K_q \sim | \oint ds \, b_k(s) \, \beta(s)^{k/2} e^{i(m\phi)+(mQ-q)2\theta} | \tag{4}$$

(m = $-k, -k + 2, \ldots k - 2, k$; θ is the machine azimuth). Certain components of this decomposition drive forces which oscillate

---

[2] The impact of the nonlinear forces on the beam depends on the β or focussing function, which is a property of the accelerator lattice and which has the machine periodicity. A large β-function means large local oscillation amplitudes but small trajectory slopes. Thus nonlinear kicks cause a relatively large distortion co m pared to positions with smaller β's. The betatron phase advance between the locations where nonlinear forces act determines whether the forces are interfering positively or negatively.

with the betatron frequency. This causes the resonant behaviour mentioned in section 2. Each of these very harmful components could be compensated by just a single pair of correction elements in the ring. However the remaining nonresonant harmonics of the nonlinear potential are also harmful if they are strong enough. These nonresonant harmonics combine to produce so called higher order resonances and detuning effects so that in principle all possible nonlinear resonances will be driven by a particular multipole component. Compensation of all of these harmonics can only be accomplished by distributed correction coils inside the magnets. A lumped corrector scheme can only compensate for parts of the whole spectrum. Nonlinear perturbation theory tells us that the contribution of a nonresonant harmonic k_q to higher order resonance driving terms or detuning terms decreases linearly with its distance from the resonant harmonic $p-mQ$ (p,m are integers, Q is the resonant tune)

$$contribution\ to\ higher\ order\ terms \sim \frac{k_q}{mQ - q} \qquad (5)$$

where we assume that the resonant harmonic has already been compensated ($k_p = 0$ for $mQ = p$). The increasing denominator describes the effect of averaging of high frequency components of the nonlinear potential.

Now we can understand how a lumped corrector scheme in particular Neuffer's scheme works. The systematic field error of the dipoles are considered constant in a half FODO cell in between the quadrupole magnets. The small spaces between dipole magnets are neglected if there is more than one dipole per half cell. The integrand of the Fourier integrals is a smooth and slowly varying function over each half cell provided that the harmonics considered is not too far from the resonant harmonic so that the change of the argument of the trigonometric function is small compared to π. The most important part of the spectrum however has that property since the impact of distant harmonics is suppressed. All integrands are then well represented by a parabola and the integral over the parabola can be expressed quite well by the values of the integrand at discrete break points with appropriate weight factors. The original Neuffer scheme uses Simpson's rule for numerical integration with 3 breakpoints. The one at the beginning and the one at the end is weighted by one respectively, the one in the middle by a factor of four. Thus Neuffer's scheme consists of a corrector next to each quadrupole in each half FODO cell which has a strength corresponding to 1/6 of the integrated multipole strength of the dipole string and one corrector in the middle which carries 4/6 of this strength.

There are many attempts to optimize or to improve lumped corrector schemes (see for example[23]), the basic idea remains the same. The question whether lumped systems are feasible cannot be answered in general, but has to be decided case by case. A general problem of all lumped schemes is its sensitivity against distortions of the closed orbit and of the beam optics.

In the case of HERA a mixed system has been chosen. Distributed coils correct the persistent current sextupoles. They consist of six meter long sextupole compensation coils which are installed inside the 9m long dipole magnets. They correct the persistent current sextupole almost perfectly, and, in addition, provide the necessary sextupole strength for chromatic corrections. The duodecapole component of the quadrupole field has to be compensated also by a correction coil inside the quadrupole magnets. Lumped correctors in the middle of each half FODO cell system compensate partly the effect of the decapoles in the dipole. They are mounted on the beam pipe inside one of the two dipole magnets in each half FODO cell and extend over 3m (see fig 5). The dynamic aperture is not completely restored by this compensation scheme, but the solution appeared to be an acceptable compromise (see fig 4).

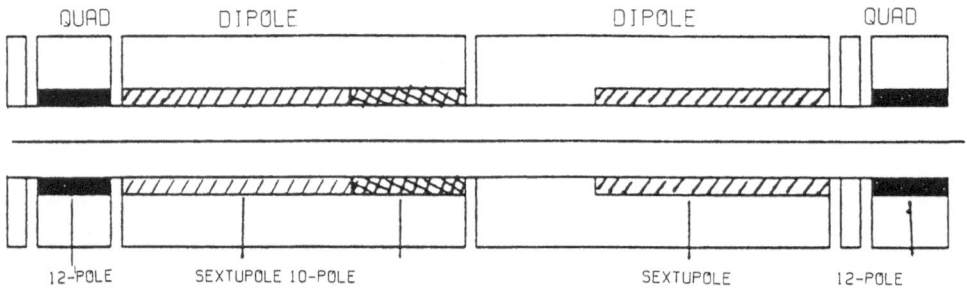

CORRECTION COILS

Figure 5. Schematic View of a HERA half FODO Cell with Correction
Coils for Persistent Current Sextupoles, Decapoles and Duodecapoles

Compensation of Nonsystematic Field Errors. Nonsystematic field errors could be compensated by using the idea of lumped compensation together with a binning procedure. Magnets with similar field errors are sorted in bins (magnet binning). In the same half cell only magnets out of the same bin are installed. Field errors within the half cell are then compensated in the same way as a systematic field error. Each half cell needs to be connected to the correction circuit corresponding to its "bin". The disadvantage of this scheme is a complicated logistics which bears many possibilities for errors and problems during installation and operation.

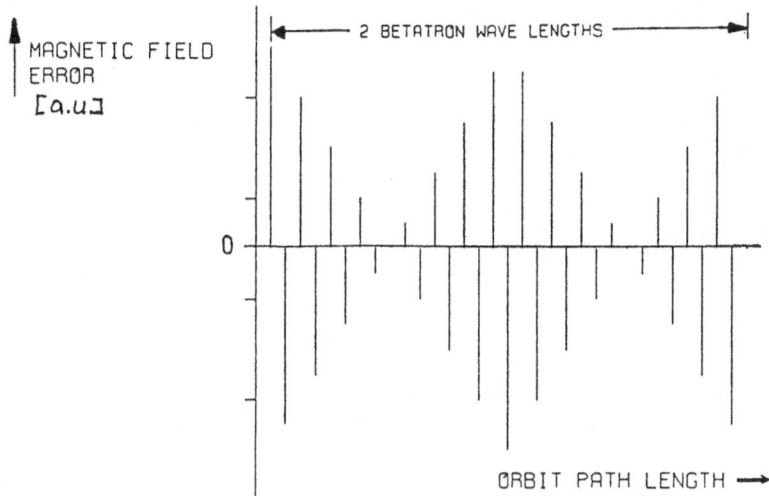

Figure 6. Example for a Sorting Pattern with a Strong Harmonic Content Near Half the Betatron Tunes

So called magnet sorting is quite inexpensive. It only requires the knowledge of the field errors. Magnets to be installed in a small section of the lattice are selected and arranged such that no resonant harmonics will be excited. The natural length of such a section is one or several betatron wave lengths. Magnet sorting is very much facilitated by making horizontal and vertical betatron phase advances exactly the same. The different resonant terms which are driven by the same multipole components[3] are then always simultaneously compensated. As far as the suppression of nonresonant harmonics is concerned the considerations are the same as for compensation of sytematic field errors. The nonlinear potential is decomposed into harmonics. Compensation of resonant harmonics is assigned highest priority. In addition, the nonresonant harmonics inside the band around the resonant harmonics are the target of the compensation scheme. A conceptually simple but nonetheless quite powerful scheme is to arrange the magnets by creating a field error pattern which has a periodicity corresponding to two betatron wavelengths[26]. In this case, the harmonic content of the nonlinear potential is reduced near integer multiples of the tunes which means it is reduced in the vicinity of the nonlinear resonances as required. The corresponding pattern of sorted magnetic field errors is sketched in fig.6. This sorting scheme has been tested successfully by simulations in the SSC[24].

---

[3] A sextupolar force $F_x \sim x^2 - z^2$, $F_z \sim xz$ drives for example the resonances $3Q_x = p$, $Q_x = p$, $Q_x \pm 2Q_z = p$.

CONCLUDING REMARKS

Dynamic aperture is an important aspect for the design of superconducting magnets. Considerable progress has been made recently in understanding and solving nonlinear problems in accelerators. Particle tracking programs which simulate the motion in accelerators have been written. New procedures to analyse the results which come out of such programs have been developed. Modern techniques are being used for a analytic description of the nonlinear motion in accelerators. The importance of complementing theoretical studies of nonlinear behaviour by machine experiments in order to determine the relevance of the results for a real machine has been recognized and first machine experiments have been carried out successfully both at CERN and at FERMILAB. This effort provides powerful tools for the design of future large accelerators.

REFERENCES

1   F.Willeke, "The Aperture of the HERA Proton Ring" Proceedings of the Second Advanced ICFA Beam Dynamics Workshop, Lugano 1988, CERN 88-04

2. B.V. Chirikov: "A Universal Instability of Many-Dimensional Oscillator Systems", Physics Reports 52, No 5 (1979)

3. A.M.Lyapunov:"Stability of Motion", Academic Press New York (1966)

4. F. Schmidt, Thesis, University of Hamburg, DESY HERA 88-02 (1988)

5. H. Mais, G. Ripken, A. Wrulich, F. Schmidt: "Particle Tracking", Lecture given in the CERN Accelerator School, Oxford 198, CERN 87-03 (1987)

6. A. Wrulich:"Particle Tracking for HERA", Accelerator Physics Issues for a Superconducting Super Collider (M. Tigner ed), Ann Arbor(1983)

7. S. Peggs:"Hadron Collider Behaviour in the Nonlinear Numerical Model Evol" Particle Accelerators, 1985, Vol 17, pp11-50

8. J. Gareyte, A Hilaire, F. Schmidt: "Dynamic Aperture and Long-Term Particle Stability in the Presence of Strong Sextupoles in the CERN SPS", CERN SPS/89-2 (AMS)

9. A.W. Chao: "Magnet Field Quality Requirements for the SSC", SSC-75 (1986)

10. see for example K.H. Meß, P.Schmüser: "Superconducting Accelerator Magnets" DESY HERA 89-01

11. see for example R. Meinke, these proceedings

12. The values listed in table 2 have been extracted from a data bank at FERMILAB containing the original magnet measurement data and which are described for example in R. Hanft et al:"Magnetic Field Properties of FERMILAB Energy Saver Dipoles", IEEE Trans Nucl Science Vol NS-30 no 4 (1983)

13. F. Willeke: "Study of the Nonlinear Acceptance of the TEVATRON by Computer Simulations", FERMILAB TM 1220 (1983)

14. N. Gelfant: "Calculation of the Dynamic Aperture of the TEVATRON" Proceedings of the Workshop on the SSC, Ann Arbor (1983)

15. F. Willeke: "Analytical Study of the TEVATRON Nonlinear Dynamics", FERMILAB FN-422 (1985)

16. R. Brinkmann, F. Willeke: "Persistent Current Field Errors and Dynamic Aperture of The HERA Proton Ring", Proceedings of the Second Advanced ICFA Beam Dynamics Workshop, Lugano 1988, CERN 88-04

17. H. Brück, R. Meinke, F. Müller, P. Schmüser: "Field Distortions from Persistent Currents in the Superconducting HERA Magnets" DESY 89-41 (1989)

18. P.W. Anderson: "Theory of Flux Creep in Hard Superconductors" Phys Rev Letters9, 309 (1962)

19. D. A. Finley, D. A. Edwards, R. W. Hanft, R. Johnson, A. D. McInturff J. Strait: "Time Dependent Chromaticity Changes in the TEVATRON", Proceedings of the 1987 Particle Accelerator Conference, Washington D.C., (1987) p151-3

20. R. Brinkmann, DESY, private communication

21. K. Johnsen, these proceedings

22. D. Neuffer: "Lumped Correction of Systematic Multipoles in in Large Synchrotrons", Proceedings of the Second Advanced ICFA Beam Dynamics Workshop, Lugano 1988, CERN 88-04

23. D. Bintinger et al: "Compensation of SSC Lattice Optics in the Presence of Dipole Field Errors", SSC-SR-1038

24. L. Schachinger, private communication

25. A. Wrulich: "RACETRACK, a Computer Code for the Simulation of Nonlinear Particle Motion in Accelerators", DESY 84-026 (1984)

26. R. Gluckstern and S. Ohnuma: "Reduction of the Sextupole Distortion by Shuffling Magnets in Small Groups", IEEE Trans on Nucl. Science Vol NS32 no p2314 (1985)

HIGH-FIELD SUPERCONDUCTING MAGNETS

FOR PARTICLE ACCELERATORS

R. PERIN

CERN, European Organization for Nuclear Research

Geneva, Switzerland

ABSTRACT

Particle Physics requires higher and higher energy ac-
celerators-colliders to explore the fundamental components of
matter. For electron-positron accelerators, circular machines,
which make use of low bending magnetic field to limit synchrotron radiation, have probably reached their maximum size
with LEP and the trend for the future is towards linear facilities. For high energy hadron accelerators/colliders, superconducting magnets have been adopted in order to save electrical
energy and reduce machine size, and the quest for cost efficiency pushes towards higher field. The 9 to 10 tesla range has
already been attained in short dipole magnets.

The large proton accelerators/colliders presently under
construction, approved or planned represent the largest scale
application of superconductivity and require massive production
in industry of advanced technology conductors, magnets and
cryogenic equipment.

A review is given of the magnets for these projects and
of the future trends supported by R.&D. programs.

INTRODUCTION

High-energy Physics requires higher and higher energy
beams to probe deeper in the structure of matter. Generally in
the past new major particle accelerators provided an order of
magnitude increase in energy with respect to their predecessors
and this trend is continuing. For protons (hadrons) accelera-
tor/colliders, synchrotrons are still the most efficient and
economical machines to accelerate store and collide high energy
beams with intensity adequate to produce the desired rate of
rare events that physicists want to study. As in circular machines the attainable beam energy is proportional to radius and
bending magnetic field, any new forward step in energy corresponds to an increase of size or field or both. The last high
energy proton machines using classical magnets (which are lim-

ited to bending fields $\leq$ 2 T) have been built more than a decade ago, and were it not for superconductivity, they would probably have been the last of the species not only because of the required large capital investment, but also of a prohibitively high electric power consumption.

Fortunately, as it sometimes happens when a technology reaches its limits, in the 1970's superconductors were developed to an adequate level and a sufficient understanding of superconducting accelerator type magnets was accumulated through the effort of several laboratories so that construction of large numbers of superconducting magnets could be envisaged. Historically the first superconducting magnets which operated reliably in an accelerator/collider were eight high gradient, large aperture, quadrupoles for the low-β insertion of the Intersecting Storage Rings (ISR) at CERN. They were soon followed by an entire gigantic (for ten years ago) machine, the Fermilab Tevatron. The importance of this machine is universally recognized: it has been for many years the largest application of superconductivity and cryogenics and a true prototype of a new generation of particle accelerators. It was (and still is) in the Tevatron that most of the unknown problems of this new technology were found and solved. A great tribute must be paid to the pioneers who had the courage to propose and the determination to successfully build it. In the meantime cryogenic plants and installations, while increasing in size, have greatly improved in efficiency and reliability, thus providing the necessary dependable environment for the safe working of the nowadays technically usable superconductors.

Superconductivity had a sixty years long infancy before being applied on a large scale to do something that could not be done without it: today and future high-energy hadron accelerators/colliders would not be built without superconducting magnets.

QUEST FOR HIGHER FIELD AND LIMITATIONS

The main motivation for higher fields is economy in capital investment and space. In addition, considering the size of presently proposed colliders, more compact machines are geographically easier to accommodate and socially more acceptable. Fig.1 shows schematically how the cost of magnets per T.m vary with field level. The plots include the cost of magnet cryostats, but the cost of the cryogenic plants is not included as it varies little with field level for the same operating temperature, the size of the plants being usually determined either by a wanted cool-down time or by beam heating. The cost increase for a 1.8 K superfluid helium system with respect to a usual 4.2 K installation is relatively small and can be estimated to be 8 to 10% of the cost of the cryo-magnetic system. Of course global machine costs profit very much from high field levels as civil engineering, infrastructure and installation costs reduce in inverse proportionality to field in first approximation. A plot of global cost vs. field would, therefore, show a much greater advantage in going to high fields. When considering running costs, this trend is dramatically reinforced. Electric power consumption, which is mainly determined by cryogenics, is in fact almost independent from field level. When comparing with the power consumption of classical magnets

the gain is striking: e.g. economies by factors as large as 60
and 100 are estimated for the LHC and the SSC respectively.

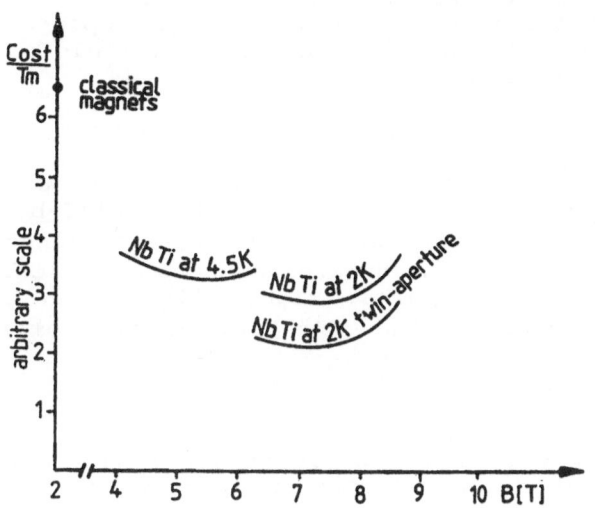

FIG.1   Cost of magnets + cryostats for 50 mm Φ coil aperture
        dipoles

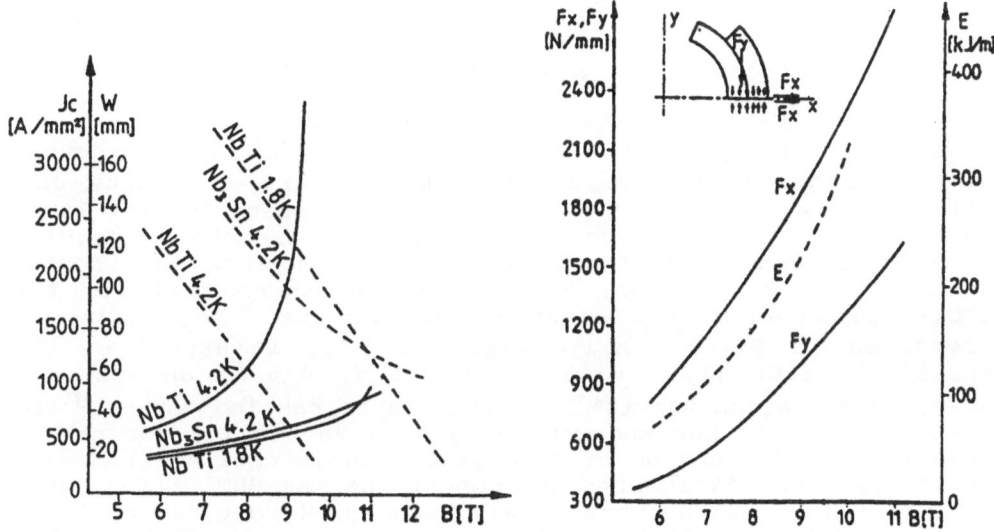

FIG.2   Current density in commercial
superconductors (wires, non-Cu part),
and coil thickness in dipole magnets
with graded current density and av.
Cu/Sc = 1.7:1

Fig.3 E.m. forces and
stored energy in 50 mm
Φ coil aperture dipole
magnets.

The way to high fields is, however, a very difficult uphill one, as illustrated in Fig.2 and 3, because of intrinsic characteristics of superconductors, rapidly rising electromagnetic forces and stored energy which severely complicate the problems of the force containment structure and magnet protection at resistive transitions. Moreover field errors at low field due to persistent currents in the superconducting filaments tend to be more important in high field magnets as they require more superconductor.

The largest accelerator/collider presently in the installation phase, the HERA[1] proton ring at DESY was designed for a 4.7 T nominal bending field, though the mass produced magnets behave so well that it is hoped that they will be run at fields well above 5 T. The UNK[2] at Serpukhov is designed for 5 T, and the largest project, the Superconducting Super Collider (SSC)[3], is based on 6.6 T dipoles. Higher fields, in the 8 to 10 T range are foreseen for the CERN Large Hadron Collider (LHC)[4].

SUPERCONDUCTORS

The only practical superconductors for magnet coils are still NbTi and Nb_3Sn. Of these two, only NbTi is produced industrially in large quantity and can be bought at competitive price. In the recent years industrial firms in many countries, stimulated by accelerator projects in particular HERA, SSC, UUK and LHC and supported by universities and laboratories have carried out a considerable effort to produce conductors which meet the requirements of particle accelerators, i.e.:

- high current density for efficient magnet design,

- small filament diameters to limit persistent currents,

- small current density spread throughout the production,

- large mechanically stable keystoned cables.

Figs 4 and 5 show the current densities of conductors recently produced in the USA and Japan for the SSC magnet development programme[5,6]. It results that a current density Jc = 2'700 A/mm^2 at 5 T, 4.2 K may at present be specified for large quantities of NbTi superconductor with an expected rejection rate not exceeding 5%. Progress has been also made in reducing Jc spread in production: a hystogram of measurements on the cables produced by one manufacturer for the HERA dipoles is presented in Fig.6[7]. All cables had a Ic in excess of the specified value (Ic = 8'000 A at 5.5 T, 4.6 K) and the I_c standard deviation was ±2%.Improvements in cabling methods[8] together with a better understanding of wire quality for cables have led to a reduction of Jc degradation in cabling from ~10% in the past to ~5% at present. Concerning the NbTi alloy, 46.5 % Ti by weight is the most widely used for 4 to 6.5 T applications. For higher field applications, e.g. in superfluid helium, an alloy richer in Ti is probably recommandable, but more optimization work has to be made. Copper to superconductor ratio has, of course, to be optimized for each magnet design and operating conditions. It is currently in the range 1.3 : 1 to 2 : 1, although many experts tend to think that 1.3 : 1 is a too low value.

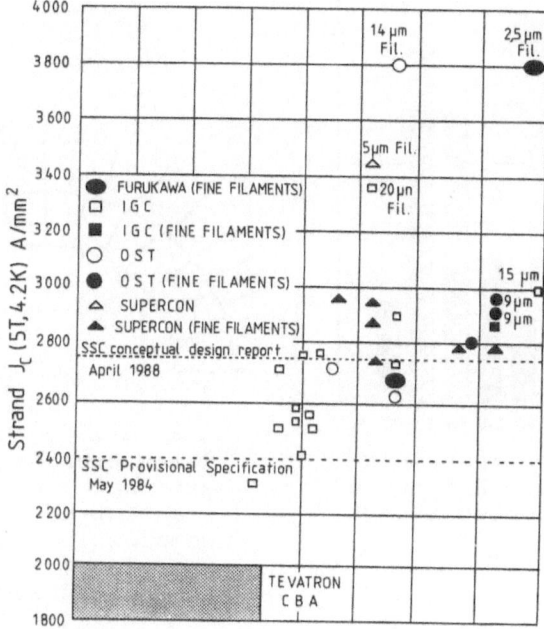

Fig.4   Current densities (4.2 K, 5 T) in NbTi wires.R&D
billets

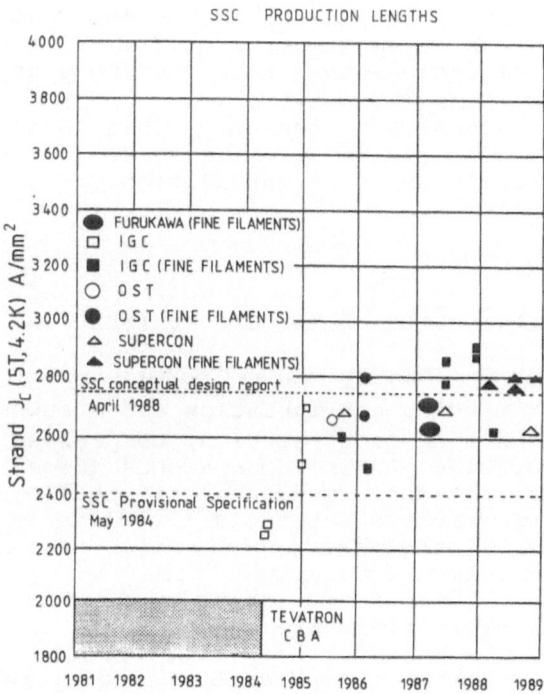

Fig.5   Current densities (4.2 K, 5T) in NbTi wires produced
for the SSC program. Production billets

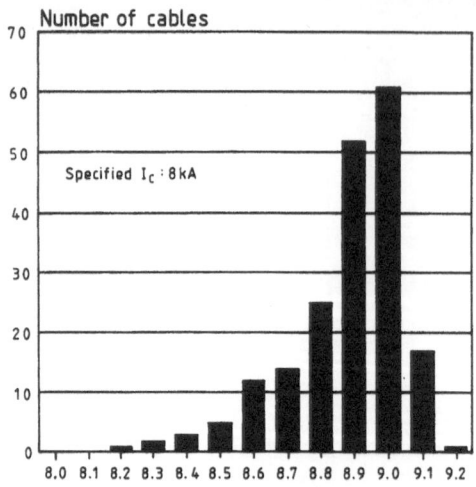

Fig.6 Critical current in kA
at 5.5 T/4.6 K in HERA
cables produced by ABB

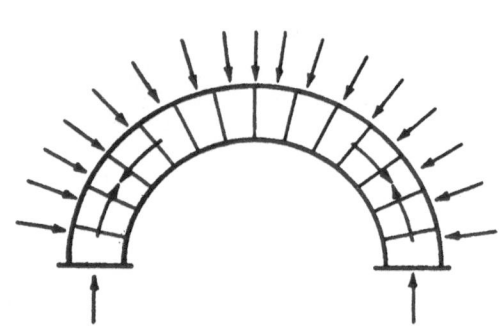

Fig.7 Roman Arch Analogy:
external pressure is
transformed into azimental
compression that prevents
tensile stresses inside
the coil structure

For very high field magnets, large aspect ratio (width/ thickness) keystoned Rutherford type cables are required. The largest ones have been developed by industry for the LHC magnet development program ($1.30/1.67 \cdot 17$ mm^2 and $2.06/2.44 \cdot 17$ mm^2). These cables have still sufficient mechanical stability and can be wound around 5.5 mm radii as required in the LHC magnet ends. A new type of cable with a keystone angle up to 3 degrees was developed in Japan[9] to allow winding of small aperture magnet coils in an ideal arch geometry. This "braid-in-strands" cable should in principle be advantageous for the mechanical behaviour of the coils and ease fabrication.

COIL CLAMPING STRUCTURES

Cos θ current distribution, Roman arch structures

All projects presently under construction or proposed adopt the "cos θ" winding configuration and a support structure which, in the transverse cross-section, compresses the coil radially from the outside. As in Roman arches (Fig.7), radial inwards compression is transformed into azimuthal pressure inside the coil structure which counteracts the formation of tensile stresses that would otherwise appear under the action of the electro-magnetic forces. At the same time the external support structure has to limit coil movements and deformation to very small values to prevent premature quenches and limit field distribution errors. To achieve this, considerable pre-loads are necessary, in particular at operation temperature. These concepts are nowadays well understood and are studied in detail by means of refined computational and experimental methods[10,11,12]

For magnets working at moderate fields, up to about 5.5 T, coil clamping is in general achieved simply by collars which are adequately compressed when mounted around the coils and locked in position by means of dowels or keys. Different solutions are however adopted by designers. E.g. when comparing the cross-sections of the HERA (Fig.8) and UNK (Fig.9) dipoles, designed for 4.67 T and 5 T nominal field respectively, important differences can be seen in the collaring system :

The material: aluminium alloy for HERA, stainless steel for UNK.
Dowel rods through circular holes to lock the collars in HERA, keys inside grooves in UNK.

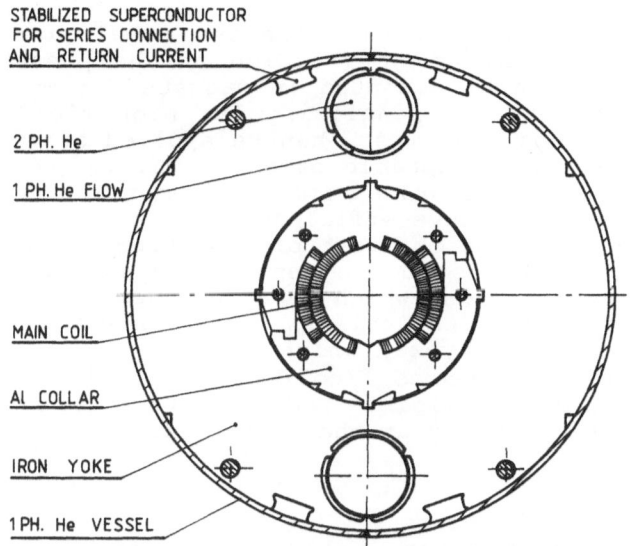

STABILIZED SUPERCONDUCTOR
FOR SERIES CONNECTION
AND RETURN CURRENT

2 PH. He

1 PH. He FLOW

MAIN COIL

Al COLLAR

IRON YOKE

1 PH. He VESSEL

Fig.8   Cross-section of HERA dipole.

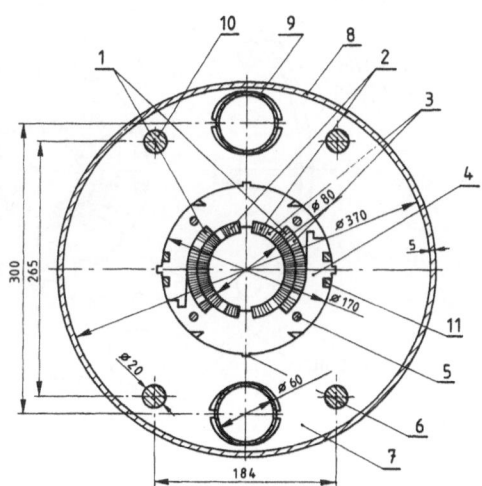

1. Outer layer
2. Inner layer
3. Spacers
4 Collars
5. Pin
6. Rectilinear lug
7. Magnetic shield
8. Helium vessel
9. Two-phase helium pipe
10. Stud
11. Key

Fig.9 Cross-section of UNK dipole.

93

The main advantage of aluminium alloy collars is that they produce an increase of the prestress on the coil at cooldown, thanks to their high thermal contraction. In this way the high prestress is applied only when needed and not at room temperature, where creep of insulation or even of copper may occur in the long run. Other advantages are lower cost of material and fabrication and the absolute absence of ferromagnetism, which is beneficial to field quality. A drawback when compared to stainless or other non-magnetic steels is that more space (wider collars) is required because of lower elastic modulus and tensile/compressive strength. In addition aluminium alloys are more sensitive to stress concentration especially under fatigue stress conditions, so that it is not advisable to use grooves and keys to lock them.

On the other hand with stainless steel collars one is bound to loose an important fraction of the prestress during cooldown . This can be accepted in magnets for moderate field levels, but is severely penalizing for high field magnets in which the prestress that must then be applied at room temperature may exceed the acceptable creep limits (magnets should be able to survive long storage time at room temperature). It is recognized that to achieve efficient designs for fields above 6 T, it is necessary that the rest of the structure (yoke + other components) contributes to the containment of the forces. So, e.g. in the HERA dipoles, which were designed for a lower operation field, above 6 T the collars come in contact in the horizontal plane with the yoke, which acts as a stopper to limit the deformation of the coil/collar assembly[13]. This however happens after a non-negligible radial outward elastic expansion (~ 0.1mm) of the assembly at the median plane.

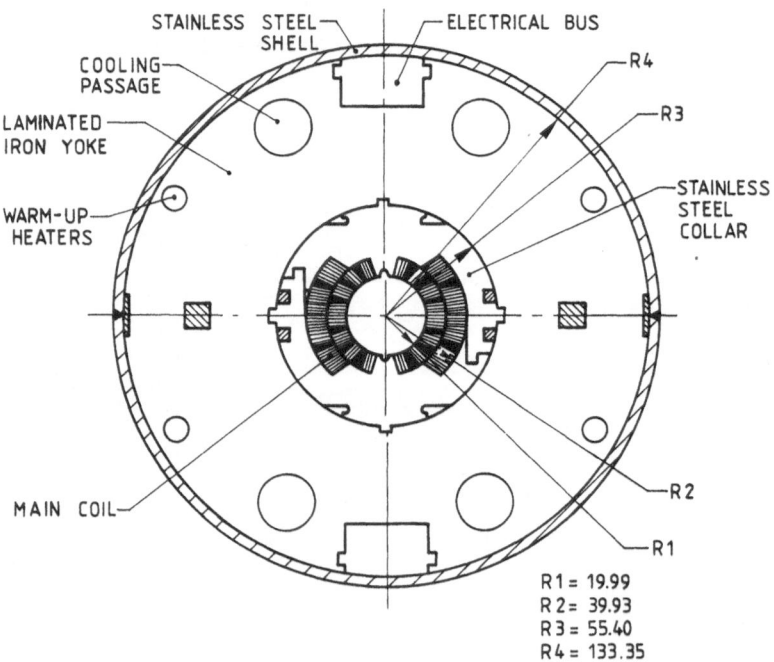

R1 = 19.99
R2 = 39.93
R3 = 55.40
R4 = 133.35

Fig.10 Cross-section of the SSC dipole.

The present design of the SSC dipoles[14] (Fig.10 includes the contribution of yoke and outer shell to the support of the windings. The horizontally split yoke fits tightly around the collars, but an important fraction (~ 30 %) of the pre-stress on the coil is lost at cooldown due to the unfavourable interplay of different thermal contraction materials. Hence the need of a higher collaring prestress at room temperature with the risk that part of it may vanish due to creep.

For the LHC 8 to 10 T dipoles a new type of mechanical structure was designed[15] in which aluminium alloy collars are surrounded by a vertically split yoke (with open gap at room temperature) clamped by a stainless steel or aluminium shrinking cylinder (Fig.11). The collars are clamped around tho coils at room temperature with a moderate pressure to avoid any risk of room temperature creep of coil components (especially of organic materials which may flow under stress at room temperature). During the cooling process the shrinkage of the collars increases the pre-compression in the coils, while the two halves of the yoke, which are actuated by the outer shrinking cylinder, move horizontally ihwards applying additional compressive forces to the collar-coil assembly in the direction just opposite to the main action of the e.m. forces . The gap between the two parts of the yoke closes at a predetermined lower temperature and when cool-down is completed a compressive force will be produced at the mating face between the two yoke halves. If this force is equal or larger than the horizontal resultant of the e.m. forces, the gap remains closed in all operating conditions and the split iron behaves as a single stiff solid body. Coil displacement and deformation under the electro-magnetic forces are, therefore, greatly reduced, as compared to a simple collar clamping system, with beneficial effects on magnet stability and also on field quality.

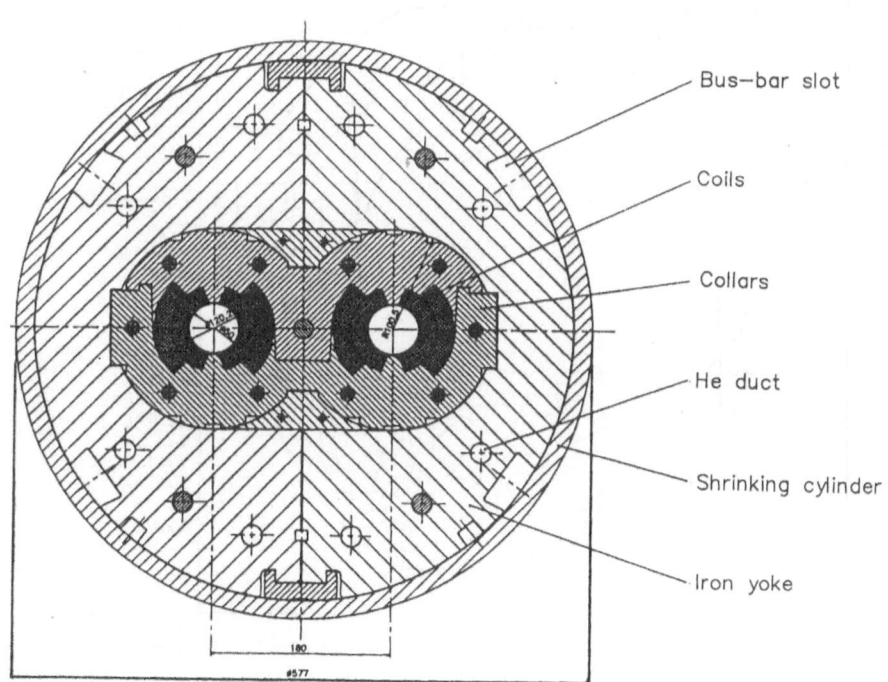

Bus—bar slot

Coils

Collars

He duct

Shrinking cylinder

Iron yoke

As an example, in the first model magnet for the LHC[16], Fig.18, the horizontal radial outward elastic displacement of the coil inner edge at the median plane at 9.3 T field was only 0.040 mm. The maximum compressive stress in the coil at 9 T central field was computed to be 11.7 dN/mm^2, which is to be compared to 18 dN/mm^2 computed for a structure of the same dimensions, but without the compressive pre-stress on the split iron mating faces. The price to be paid for this more efficient, but also more complex, structure is that its components have to be produced with tighter dimensional accuracy to ensure that they perfectly fit together.

A problem with present coil collaring systems is that the coils have to be compressed more than strictly necessary during collaring in order to produce the clearances (typically 0.1 mm) which permit the insertion of dowel rods or keys and to compensate the elongation of the collar "legs" under the reaction of the coils when releasing the press. Both effects compel to overstress the coil for a short while at collaring and this is particularly harmful for high field magnets. Fig.12 shows these effects as computed for the collars/coil assembly of the LHC dipole[12]. It can be seen that, if special correcting steps are not taken, a gap of about 0.2 mm appears between upper and lower collars at the release of the press. The corresponding additional pressure to be applied to the coils under the press would be about 7 dN/mm^2. A partial solution has been found for the SSC magnet with the adoption of "tapered" keys[14]. For dowel rod type of assemblies, solutions are being sought.

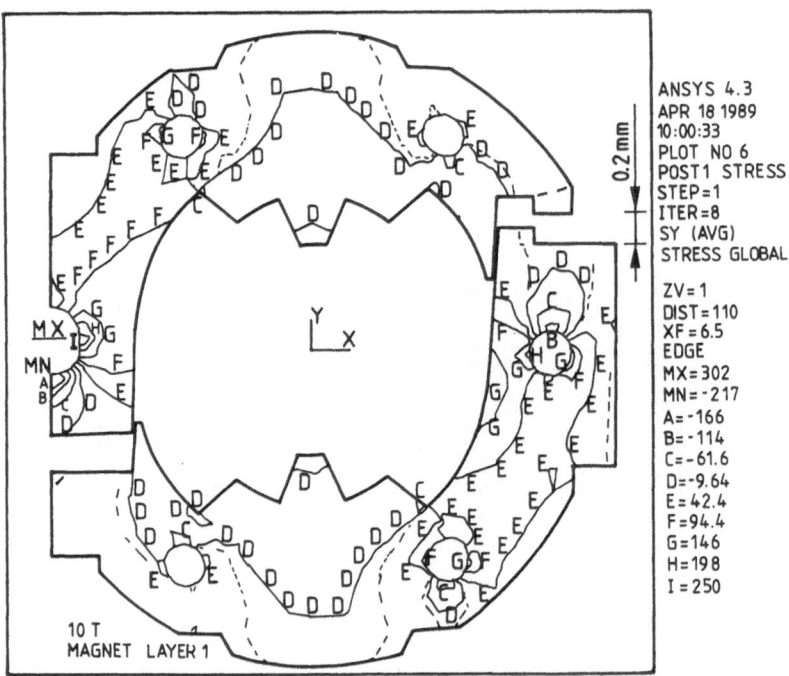

Fig.12 Computed deformations and azimuthal stress distribution in the collars of the LHC dipole after releasing the collaring press.

Axial electro-magnetic forces in high field magnets produce large stresses and strains in the coil. E.g. in the LHC 10 T,10 m long, dipole the axial tensile stress in the coils would be 130 MPa and the elastic elongation 11 mm. Magnet ends have, therefore, to be adequately supported by the external structure and a detailed stress analysis has to be done for any end design.

Other configurations and structures

An interesting configuration is the one adopted for a window frame, 1 m long, magnet built at KEK[17] which reached 9.3 T central field at 1.8 K. (Fig.17). The coils, eight of which were wound in race-track double pancakes, are assembled on inner supports and clamped with wedge supports and outer collars. Inner supports and outer wedge type supports are made of non-magnetic high manganese steel (32 Mn 7 Cr) of low thermal contraction and the collars of stainless. The interplay of these different thermal contraction materials was used to increase the compressive prestress on the winding at cooldown.

An alternative approach is the superferric concept pursued at the Texas Accelerator Center (TAC)[18]. Fig.13 shows the cross-section of a short (~0.6 m long) magnet model which reached 6.1 T. Another model designed for 8 T field has been recently built, but did not reach the design field. These magnets, however, require subdivision of the coil in at least two parts excited at different currents which need to be programmed all along the excitation cycle, to achieve an acceptable field quality.

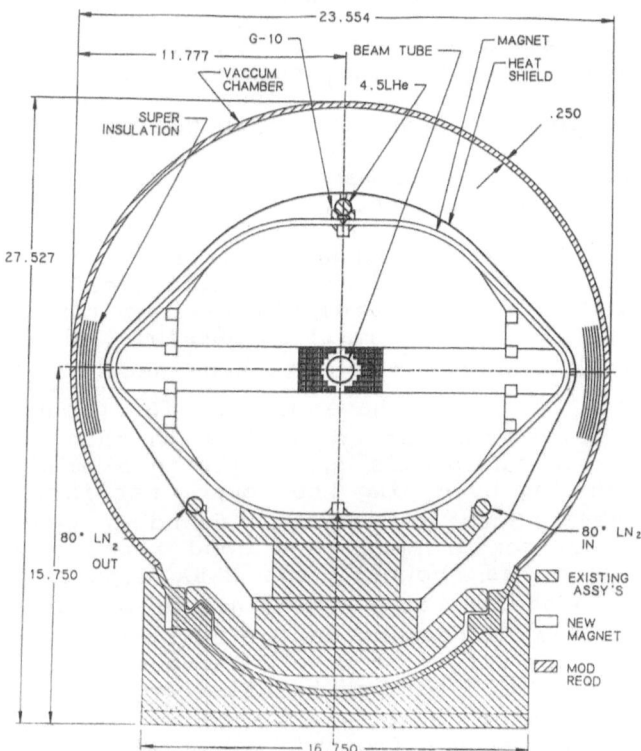

Fig.13   Cross section of the TAC 6 T model magnet.

FIELD QUALITY

Field errors are usually expressed under the form of multipole expansion components of the transverse field in two dimensions :

$B_y + iB_x = B_0 \sum_n (b_n + ia_n) (Z/R_r)^{n-1}$ where $Z = x + iy$; $R_r = $ ref.radius;

B_0 = magnitude of dipole field in the y (vertical) direction;

b_n = normal multipole coefficient; a = skew multipole coefficient;

$b_1 = 1$; a_1 = dipole x-component; a_2, b_2 = quadrupole components etc.

If one takes R_r to be about 60% of the coil inner radius, tolerances for dipole magnets, which may vary from machine to machine, are usually of the order of 5×10^{-4} for the sextupole (b_3) component and 1×10^{-4} for the higher multipoles. In the quadrupoles the corresponding limit is in general of a few 10^{-4}. Field errors are of different origin:
 - Geometrical errors in the coils and structure.
 - Deformations under the action of e.m. forces.
 - Saturation of iron yoke at high field.
 - Persistent currents at low field.

Measurements taken on several model dipoles for the SSC and RHIC[19] and on the HERA magnets show that "geometric" multipoles can be kept within the tolerance limits. Torsion along the magnet length, which produces local tilts of the median plane, can also be kept under control.

Saturation at high field, a systematic effect provided the spread of the magnetic characteristic of the yoke iron is well controlled over the entire production, can be compensated by correcting elements such as distributed correcting windings or by separate correctors. When quadrupoles are powered in series with the dipoles, mismatching due to saturation must be corrected by re-tuning elements.

Persistent currents in the superconductor are still the most worrying source of error at injection field where the beam has in general to stay for relatively long times. The first cure is to use filaments as small as possible. While conductors with 4.5 to 6 μm diameter filaments are industrially produced, 2.5 μm filament conductors have been so far produced only experimentally. More work has to be done in this domain, especially in light of the promising results in eliminating proximity effect coupling by doping the copper matrix with 0.5 percent manganese[20]. Good agreement is found in general between computations and measurements and good reproducibility from magnet to magnet, as shown by the HERA dipoles and quadrupoles[21], Figs 14 and 15. Time dependence of persistent currents effects, which was discovered on the Tevatron[22] is also found in the HERA magnets with long time constants for all multipoles including the dipole. The most harmful component, the 6-pole, presents variations of -2.5×10^{-4} in magnets from a manufacturer and -4.8×10^{-4} in magnets from the other manufacturer in 2'000 s, the time approximately spent at injection field[23].

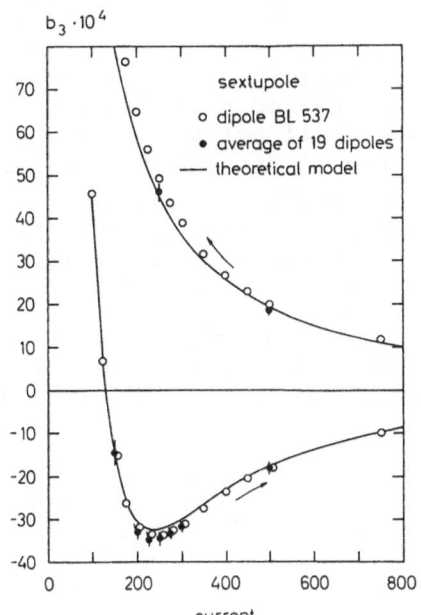

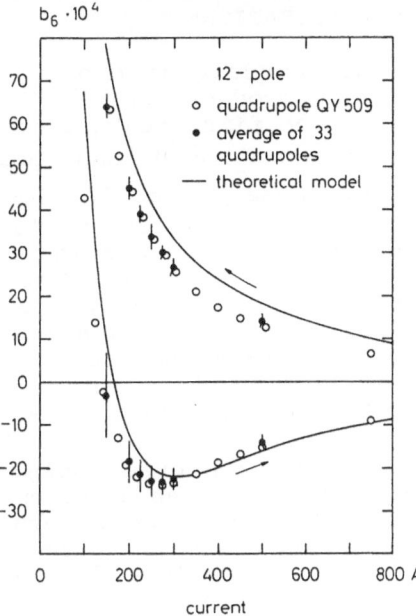

FIG.14 Current dependence of 6-pole component in HERA dipoles at low excitation

FIG.15 Current dependence of 12-pole component in HERA quadrupoles at low excitation

The phenomenon has not yet found a complete satisfactory explanation. The envisaged solution for HERA is a continuous measurement of the field in reference magnets from both manufacturers and a programmed excitation of the correcting elements to compensate the time variable errors.

QUENCH PROTECTION

Magnets have to be protected against overheating (or even burning) in case of accidental quench, which may be induced, e.g. by a beam loss. Purely passive systems, based e.g. on subdivision by means of diodes or resistors, are absolutely reliable, but unfortunately can be used only for relatively low stored energy magnets. In high field magnets with coil aperture $\geq$ 50 mm and a normal length for accelerator use (6 + 17 m), an active system making use of heaters which rapidly propagate the quench to most of the conductor mass is necessary to ensure survival of the magnets. E.g. the 8 to 10 T magnets under study for the LHC would not be able to withstand quenches without an absolutely dependable heater system. In addition, means of by-passing the current (e.g. cold or warm diodes) must be provided. A good example of protection system presently being implemented is given by HERA[24]. At a quench of a magnet, the following actions take place: cold diodes by-pass the current, quench is detected and signal amplified by means of magnetic amplifiers, heaters are fired, switches that normally short serial dump resistors are opened. In addition, groups of dipoles/quadrupoles are connected to bridges that can command firing of heaters through a control software system.

INDUSTRIALIZATION

Superconducting magnets for accelerators have become an industrial product, which satisfies the needed high standard of quality. The first to be produced in a small series in industry had been the 9 low-β quadrupoles for the ISR in the 1970's followed by those of the LEP low-β insertions[25], but the best example of industrialization is provided by the HERA proton ring: all magnets of this machine have been or are being produced by industries. So far more than half of the 453 dipoles have been built in two fabrication lines in Germany and Italy, the completion is expected in summer 1990. Excellent reproducibility has been achieved for each of the lines with a systematic 7 mm (8×10^{-4}) difference in magnetic length between magnets coming from the two production lines.

All the 246 quadrupoles (developed at CEA Saclay)[26] are completed, 126 have been manufactured in France and 120 in Germany. Also for the quadrupoles reproducibility from magnet to magnet of the same production line is very good, with a systematic 1‰ difference in integrated gradient vs. current between the two lines.

All correction elements have been satisfactorily manufactured in industry in Germany and the Netherlands.

For the SSC project all short and long prototype magnets have so far been built in laboratories (BNL, Fermilab, LBL) and an industrialization program is starting.

For the LHC, collaboration with industry has started right from the beginning of R&D. with staff from firms joining the design team at CERN, acquiring the specific know-how while at the same time helping determining industrial production processes and designing the tooling. All models and prototypes are built in industry.

RESEARCH AND DEVELOPMENT TOWARDS HIGHER FIELD

High field magnet models aimed at the 8 to 10 T field range have been developed in some laboratories. A considerable effort was carried out in the ·U.S.A. especially at LBL, where in 1983 a 50 mm aperture NbTi model attained 9.1 T at 1.8 K[27] (Fig.16). In Japan a sustained activity is going on at KEK. In 1985 a 60 mm aperture, 1 m long, NbTi model already mentioned above (Fig.17) reached 9.3 T at 1.8 K[17].

The European effort in this domain is at present concentrated on the development of high field magnets for the LHC. Several laboratories and many industrial firms collaborate very actively to this effort[28].

So far three single aperture, 50 mm coil inner diameter, 1.35 m long magnets have been built and tested. The first magnet (Fig.18) was designed for 8 T nominal field using NbTi superconductor for 1.8 K operation, in the frame of a collaboration between CERN and the Italian firm ANSALDO[16]. It had its first quench at 8.55 T central field and passed 9 T at the third quench reaching 9.3 T during three test compaigns, two at CERN and one at CEA Saclay (Fig.19). A second magnet of the same design, but using a mechanically stronger conductor insulation in order to ease future mass production of the coils, was tested in July 1989 and behaved practically as the first one, reaching 9.45 T (10660 A) maximum central field at 1.8 K.

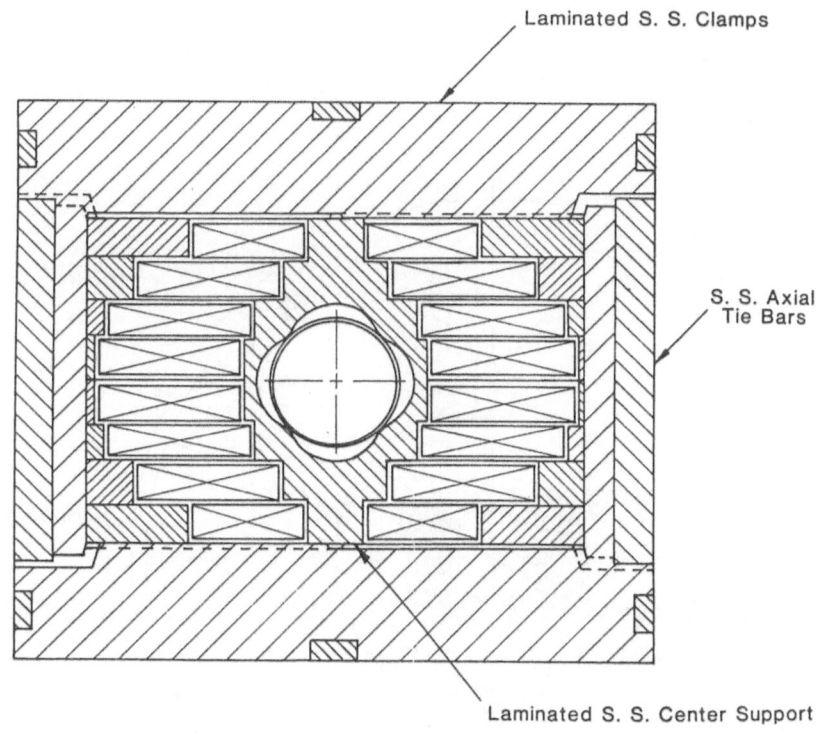

FIG.16   Cross-section of LBL high-field model magnet with pancake windings.

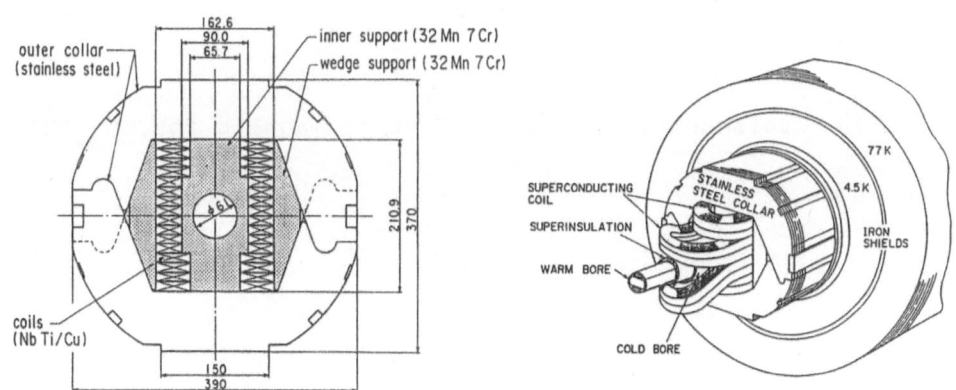

FIG.17   Cross-section and end structure of the KEK high-field model magnet.

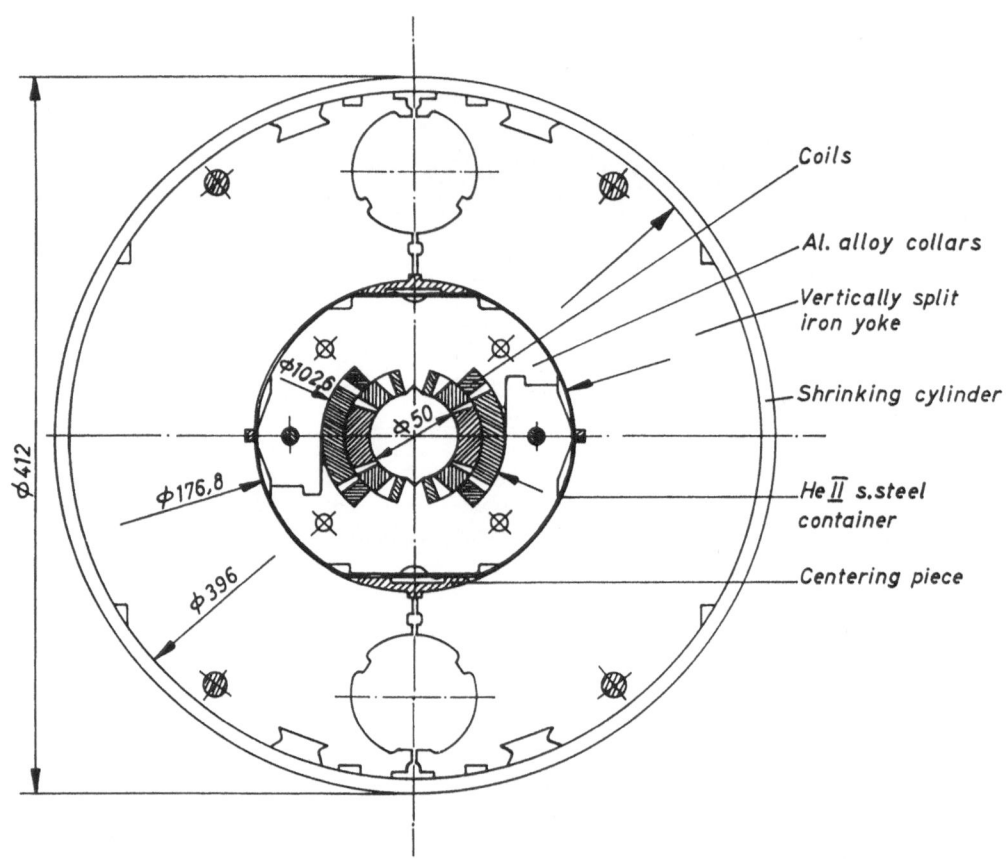

FIG.18 Cross-section of the 8 T model magnets for the LHC.

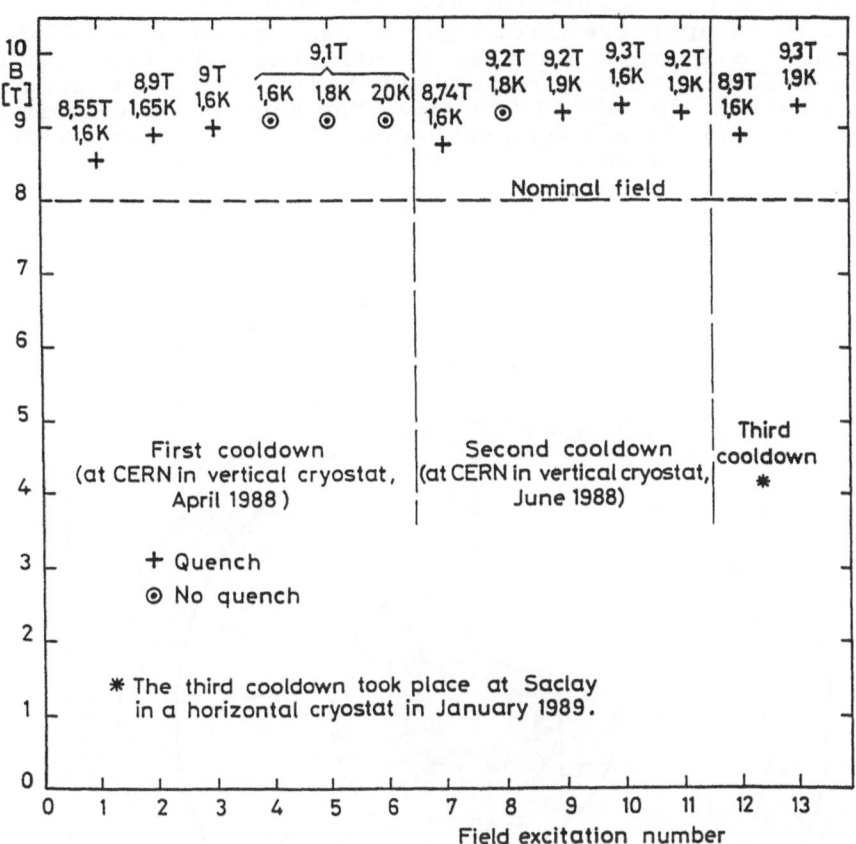

FIG.19 Quench history of the first 8 T model for the LHC.

The third model magnet used Nb3Sn conductors and was built by ELIN-UNION, Austria, in the frame of a collaboration[29] with CERN. Its cross-section is shown in Fig. 20. It was tested in June 1989 and reached 9.5 T at 4.3 K after a few quenches. It had been preceded by the test of a single coil in a magnetic mirror test device, which attained a 10.2 T field (Fig.21).

In the meantime four twin-aperture (Fig. 11), 1.35 m long model magnets[30], designed for 10 T field and using NbTi conductors for 1.8 K operation, have been ordered to four firms in Austria, France, Italy and The Netherlands. It is hoped that they will be tested in 1990. To explore further the Nb3Sn route, the development of a 10 T twin-aperture model magnet has been started in the Netherlands by a CERN-FOM-UT-NIKHEF Collaboration[31]. In parallel with the short model magnets, a 9 m long twin-aperture prototype[32] was designed at CERN and ordered to a firm in Germany. It will use HERA type coils in order to save time and money, but the rest of the structure corresponds to the LHC 10 T design. The magnet active part will be mounted inside a horizontal cryostat by a firm in Italy. Tests will be carried out at CEN, Saclay, in 1990.

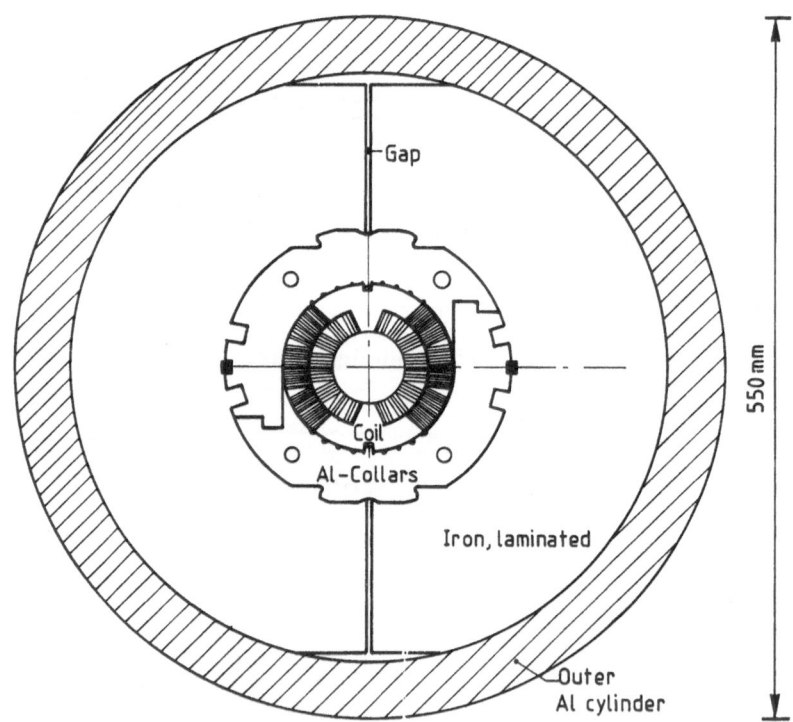

FIG.20   Schematic cross-section of the 1 m long Nb3Sn mode magnet for the LHC.

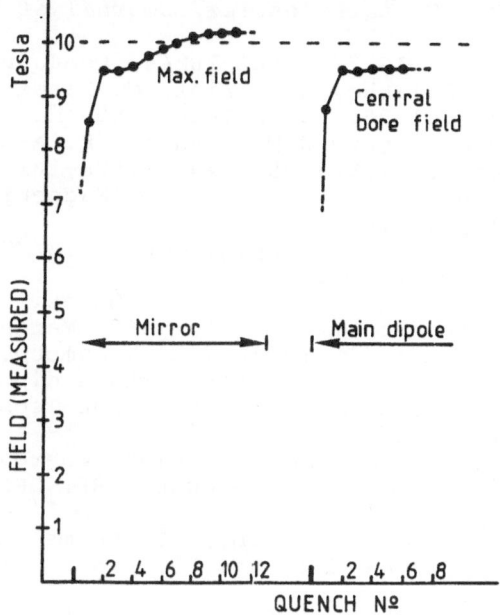

FIG.21   Quench history of the Nb3Sn mirror dipole and of the full dipole.

CONCLUSION

The technology of superconducting magnets for fields up to 6.5 T and based on NbTi superconductors operating at 4.2 + 4.5 K is well developed and applied on a large scale. Improvements are still needed, in particular on the reduction of persistent current effects. For higher fields the NbTi technology at superfluid helium temperature requires further experimental studies concerning the cooling of long strings of magnets and thermal stability under beam loss conditions, especially for the proposed high intensity colliders. The Nb3Sn technique has the advantage of proved cooling systems and more confortable temperature margin, but has still to be tested on long magnets and is at present more expensive.

ACKNOWLEDGMENTS

The author is grateful to all colleagues who provided him with information.

REFERENCES

1   B.H. Wiik, IEEE Trans.Nucl.Sc, vol.NS-32, pages 1587-1591, October 1985.
2.   A.I. Ageev et al., Proc.1988 EPAC Conf., pages 233-235.
3.   Conceptual Design of the Superconducting Super Collider, SSC SR-2020, March 1986.
4.   The LHC Working Group, CERN 87-05, May 1987.

5. P.J. Lee and D.C. Larbalestier, Journal of Materials Sc. 23, 1988.

6. R. Scanlan, SSC-MAG-226, LBL-26244, February 1989.

7. R.K. Maix, D. Salathe, S.L. Wipf, M. Garber, IEEE Trans. on Magnetics, Vol.25, pages 1656-1659, March 1989.

8. J. Grisel, J.M. Royet, R.M. Scanlan, R. Armer, IEEE Trans. on Magnetics,Vol.25, pages 1608-1610, March 1989.

9. K. Ishibashi et al., IEEE Trans. on Magnetics, Vol.25, pages 1628-1631,March 1989.

10. M. Bona, G. Sborchia, G. Spigo, CERN/TIS/MC/03-88, SPS/EMA/88-01,January 1988.

11. C.L. Goodzeit, M.D. Anerella, G.L. Ganetis, IEEE Trans. on Magnetics,Vol.25, pages 1463-1468, March 1989.

12. D. Leroy, R. Perin, D. Perini, A. Yamamoto. Structural analysis of the LHC 10 T twin-aperture dipole. Paper presented at 11th Int. Conf. on Magnet Techn., Tsukuba (Japan), August 1989.

13. H. Kaiser, 13th Int. Conf. on High Energy Accelerators, Novosibirsk,August 1986, DESY HERA 1986-14.

14. J. Peoples for BNL, Fermilab., LBL teams and SSC CDG, IEEE Trans. on Magnetics, Vol.25, pages 1444-1450, March 1989.

15. R. Perin, CERN LHC Note 32, August 1985.

16. R. Perin, D. Leroy, G. Spigo, IEEE Trans. on Magnetics, Vol.25, pages 1632-1635, March 1989.

17. T. Shintomi et al., IEEE Trans. on Nucl.Sc., Vol.NS 32, pages 3719-3721, October 1985.

18. J.C. Colvin et al., Nucl.Instr. and Methods in Phys., Res.A270 (1988),207-211.

19. P. Dahl et al., IEEE Trans. on Magnetics, Vol.24, pages 723-725, March 1988.

20. E. Gregory et al., IEEE Trans. on Magnetics, Vol.25, pages 1926-1929, March 1989.

21. H. Bruck, R. Meinke, F. Muller, P. Schmuser, DESY 89-041, March 1989.

22. D. Finley, D.A. Edwards, R.W. Ranft, R. Johnson, A.D.McInturff,J. Strait, FNAL FN-451, March 1987.

23. H. Barton et al, and R. Meinke, Performance of the Superc. Magnets for the HERA Accel., presented at this Conference.

24. K.H. Mess, DESY HERA 87-10, April 1987.

25. Ph. Lebrun, S. Pichler, T.M. Taylor, T. Tortschanoff, L.Walckiers, IEEE Trans. on Magn., Vol.24, pages 1361-1364, March 1988.

26. R. Anzolle, J. Perot, J.M. Rifflet, A. Fokken, O. Peters, S. Wolff, IEEE Trans. on Magnetics, Vol.25, pages 1660-1662, March 1989.

27. W. Hassenzahl, G. Gilbert, C. Taylor, R. Meuser, Proc. MT8, Grenoble, pages C1-271-277, 1983.

28. R. Perin, IEEE Trans. on Magn., Vol.24, pages 734-740, March 1988.

29. A. Asner, R. Perin, S. Wenger, F. Zerobin, Paper presented at MT-11, Tsukuba, 28.8.-1.9.1989.

30. D. Leroy, R. Perin, G. de Rijk, W. Thomi, IREE Trans. on Magnetics, Vol. 24, pages 1373-1376, March 1988.

31. H.H.J.ten Kate et al., Paper presented at MT-11, Tsukuba, 28.8-1.9.1989.

32. Ph. Lebrun, D. Leroy, R. Perin, J. Vlogaert, A Mcinturff, Paper presented at MT-11, Tsukuba 28.8.-1.9.1989.

THREE-DIMENSIONAL COMPUTATION OF MAGNETIC FIELDS

AND LORENTZ FORCES OF AN LHC DIPOLE MAGNET

C. Daum

NIKHEF-H

P.O. Box 41882, 1009 DB Amsterdam, Netherlands

INTRODUCTION

For the design of magnets, a detailed knowledge of fields and forces is needed as well in the straight sections of the coil as in the coil heads. The method of computation presented here is designed for structures with shell coils around a cylindrical aperture surrounded by a cylindrical iron yoke. The effect of the iron is taken into account using the method of image currents for a fixed value of the permeability, and, hence, the variation of the permeability in the iron is not taken into account. The fields and forces are calculated as the sum of the fields and forces of the strands out of which the conductors are composed. The strands are "ideal" strands, i.e. they are parallel to the axis of the conductor and thus do not follow the actual layout of the strands in a Rutherford cable. The current is concentrated in the centre of the strands. The magnetic field due to a single strand is calculated with the Biot-Savart law using delta functions for the radial and angular current distributions. The integrals in the Biot-Savart law can now be evaluated. Fields are always calculated as the sum of the contributions of the individual strands. A detailed description is used for constant perimeter coil heads. Lorentz forces are calculated at the centre of either the strands, or the conductors, or the blocks with the current concentrated at the centre. Results on magnetic fields and field integrals, a multipole expansion of the field integrals, and the magnetic length are presented. An extensive account of the method is given in Ref.3.

AN LHC DIPOLE MAGNET

The shell coil of the prototype LHC magnet[2,3] consists of two layers. A quadrant of the coil is shown in Fig.1, which also defines the coordinate system used. The inner layer has four blocks of conductors with four, four, three and two conductors, the outer layer has two blocks of conductors with seven and seventeen conductors The conductors in the inner layer consist of two rows of thirteen "ideal" strands, those in the outer layer have two rows of twenty "ideal" strands. Both conductors have a slight keystone angle.

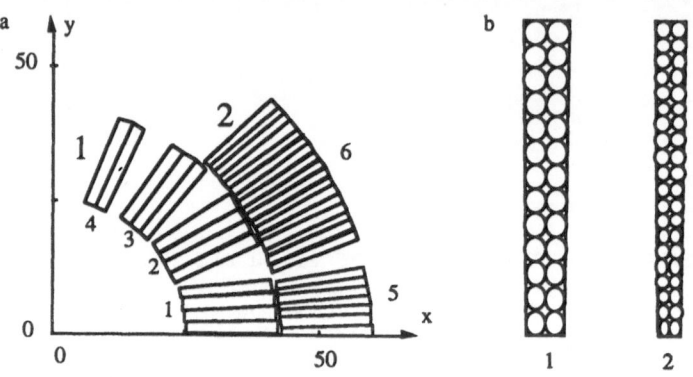

Fig.1.Layout a) of the conductors in the 6 current blocks of an LHC dipole magnet with dimensions in mm, and b) of the "ideal" strands in the conductors of layer 1 and 2

Fig.1 also shows a cross section of both conductors, which have a height of 17 mm, and have a width at top and bottom of.2.64 (1.81)mm and 2.18(1.44)mm for layer 1(2), respectively.

The LHC prototype magnet is a twin aperture magnet. The method of computation described in this paper is for a single aperture magnet with a cylindrically symmetric configuration. The shape of the "ideal" strands in the conductors is a race track with straight parts with individual length for each strand and with coil heads, which are half circles, if exposed in a flat plane, at the inner radius of the conductor, and half ellipses elsewhere for obtaining constant perimeter coil ends. The coordinate system is defined with the x-axis in the horizontal plane transverse to the symmetry axis of the magnet, the y-axis along the field direction and the z-axis along the symmetry axis.

COMPUTATION OF FIELDS, FIELD INTEGRALS AND FORCES

The basic integrals of the Biot-Savart law are expressed in a cylindrical coordinate system with the z-axis as symmetry axis of the cylinder. They can be evaluated analytically for the straight parts of each strand, if the current is taken to be concentrated at the centre of the "ideal" strands of Fig.1b. The integrals over the angular variables can be performed immediately using a delta function for the radial and azimuthal distributions at the position of the strand. The integration over z can be performed analytically for the exact length of the straight part of the strand. In the coil heads, the integration over the radial and angular variables can again be performed analytically. The integration over z is made using the Simpson rule, for which it is sufficient to make only 10 steps over a coil head. The field at any field point is obtained by summation over the contributions of all strands[1]. In the same way, also field integrals are calculated.

The iron yoke has a radius r' in the central part of the magnet and jumps to a radius r'' in the outer part before the straight parts of the coils end. These radii are used for the calculation of the position of the image current of each strand and their contribution to the magnetic field as for the direct contributions of the straight parts of the strands. The following approximations are made. It is assumed that the iron yoke extends to infinity in the z-direction as well as in the radial direction outwards from radius r' and r'' separately. Images with respect to the plane transverse to the z-axis where the actual iron yoke changes from inner radius r' to r'' are neglected. The relative permeability of the iron is taken to be $\mu = \infty$.

A comparison[1] with three-dimensional field calculations using TOSCA[4] and with the multipoles of a two-dimensional calculation using POISSON[5] shows that these approximations can be made in the case that the radii r' and r'' are sufficiently large, and hence the contribution of the image currents to the field is small.

The Lorentz force on a point of a strand is the vector product of the strand current and the field at this point due to all other strands. It has been checked[1] that the force on a conductor, taking the current to be concentrated at the centre of the conductor, is within a few percent equal to the sum of the forces on the strands of the conductor. The forces on the conductors are used in the mechanical design of the magnet. In all cases, the field and field integrals are calculated from the individual strands.

RESULTS FOR AN LHC MAGNET

A program has been written for field and force computations with the method presented above. It has been applied to an LHC magnet of a nominal length of 1m with the conductor layout of Fig.1. The field and force configuration has been calculated for a central field B(0,0,0) = 10T for which an excitation current I = 14375A is needed. Then, the current in the strands of layer 1 and layer 2 are I_{1s} = 552.9A and I_{2s} = 359.4A, respectively. The corresponding current densities are about J_{1s} = 404A/mm^2 and J_{2s} = 558A/mm^2, respectively. A graded current density is used for optimization of the current carrying capability of available conductors for the two coil layers. The iron yoke has an inner radius r'=0.100m between z = −0.302m and z = 0.302m. Outside this range, it has an inner radius r" = 0.128m. The contribution of the image currents in the iron yoke is less than 15% due to the large inner radius of the iron.

An important design criterion is the maximum field on the conductors. This occurs in layer 1 at the inner top edge of block 4, and in layer 2 close to the inner top edge of block 6 (see Fig. 1). The maximum fields in layer 1 and layer 2 are B_{1max} = 10.217T, and B_{2max} = 8.772T at z_1 = 0m, and z_2 = 0.36m, respectively. In layer 2 the field at z = 0 equals 8.591T. These maxima should be compared with the properties of available superconductors at the required current densities. The magnetic length of this configuration is L_{magn} = 0.867m.

The multipole expansion of the field integrals is

$$\int_{-\infty}^{\infty} B(x,0,z)\,dz = \left(\int_{-\infty}^{\infty} B(0,0,z)\,dz \right) \sum_{n=1}^{\infty} b_n \left(\frac{x}{x_0}\right)^{n-1} \tag{1}$$

Table 1 lists the coefficients b_n for a 1m LHC magnet with the configuration of Fig.1.

For a further detailed understanding, Fig.2 shows the direct contributions of the coils to the field on the z-axis, and to the multipole coefficients

$$c_n = \left(a_n^2 + b_n^2 \right)^{\frac{1}{2}} \tag{2}$$

of the field integrals, and those of their mirror images in the iron yoke for this magnet. Here, the coefficients a_n are the skew multipole coefficients which vanish for the used dipole symmetry. We see that all multipole coefficients due to the mirror images for $n = 5$ and larger are negligibly small.

Fig.3 shows the components of the Lorentz force as a function of the azimuthal angle θ of the conductors in the straight part of the coils at $z = 0$ and 40 cm, respectively.

TABLE 1. Multipole coefficients b_n of the series expansion (1) of the field integrals for a 1m LHC magnet; T, S and H are for Total field integral, and the contributions of the Straight part of the coil and the coil Heads, respectively, $x_0 = 20$ mm.

n	b_n (T)	b_n (S)	b_n (H)	n	b_n (T)	b_n (S)	b_n (H)
1	$1.00*10^{+0}$	$8.60*10^{-1}$	$1.39*10^{-1}$				
3	$9.17*10^{-3}$	$1.07*10^{-2}$	$-1.50*10^{-3}$	13	$-2.51*10^{-5}$	$1.41*10^{-5}$	$-3.92*10^{-5}$
5	$-5.87*10^{-4}$	$4.18*10^{-4}$	$-1.01*10^{-3}$	15	$9.88*10^{-5}$	$1.02*10^{-4}$	$-3.16*10^{-6}$
7	$1.07*10^{-3}$	$1.05*10^{-3}$	$2.48*10^{-5}$	17	$-2.16*10^{-4}$	$-1.96*10^{-4}$	$-2.07*10^{-5}$
9	$6.40*10^{-4}$	$6.37*10^{-4}$	$3.75*10^{-6}$	19	$-4.34*10^{-5}$	$-6.37*10^{-5}$	$2.03*10^{-5}$
11	$4.88*10^{-4}$	$4.91*10^{-4}$	$3.01*10^{-6}$	21	$3.60*10^{-5}$	$4.66*10^{-5}$	$-1.06*10^{-5}$

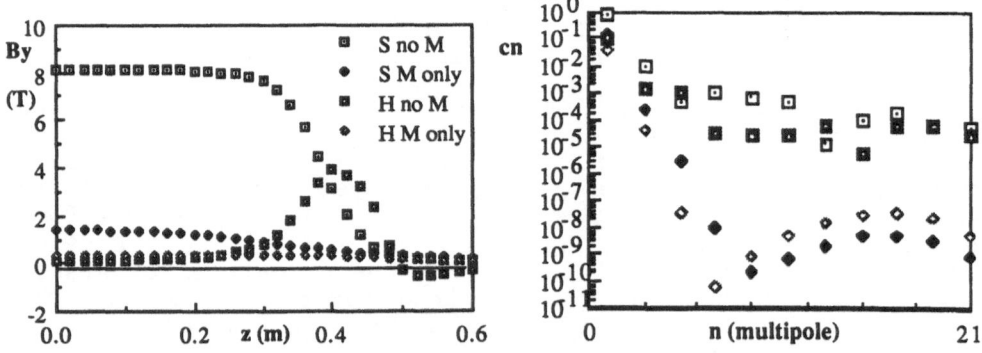

Fig.2.Field along z-axis and multipole coefficients for direct contribution of the straight parts of the strands (S no M) and the coil heads (H no M), and for their mirror contributions (S M only and H M only)

The radial force is always outward at z = 0 cm. The reversal of the radial forces at z = 40 cm within blocks needs attention in the mechanical design of a magnet. The overall radial force of the two layers is outward, which implies, however, a compression of the insulation between inner and outer layer. The azimuthal force is always towards the median plane. Hence, all forces are contained within the arch of the coil, and no forces are exerted on the beam pipe. The accumulated azimuthal force towards the median plane causes a very high stress on the conductors in the median plane. The azimuthal stress for layer 1 and layer 2 are about σ_{1a} = 57 MPa, and σ_{2a} = 70 MPa, respectively. The azimuthal forces require a high azimuthal prestress, which add to the stress values, and can not be chosen to be a comfortable factor of more than two higher than the azimuthal stresses at maximum excitation as in the case of the HERA magnets[6].

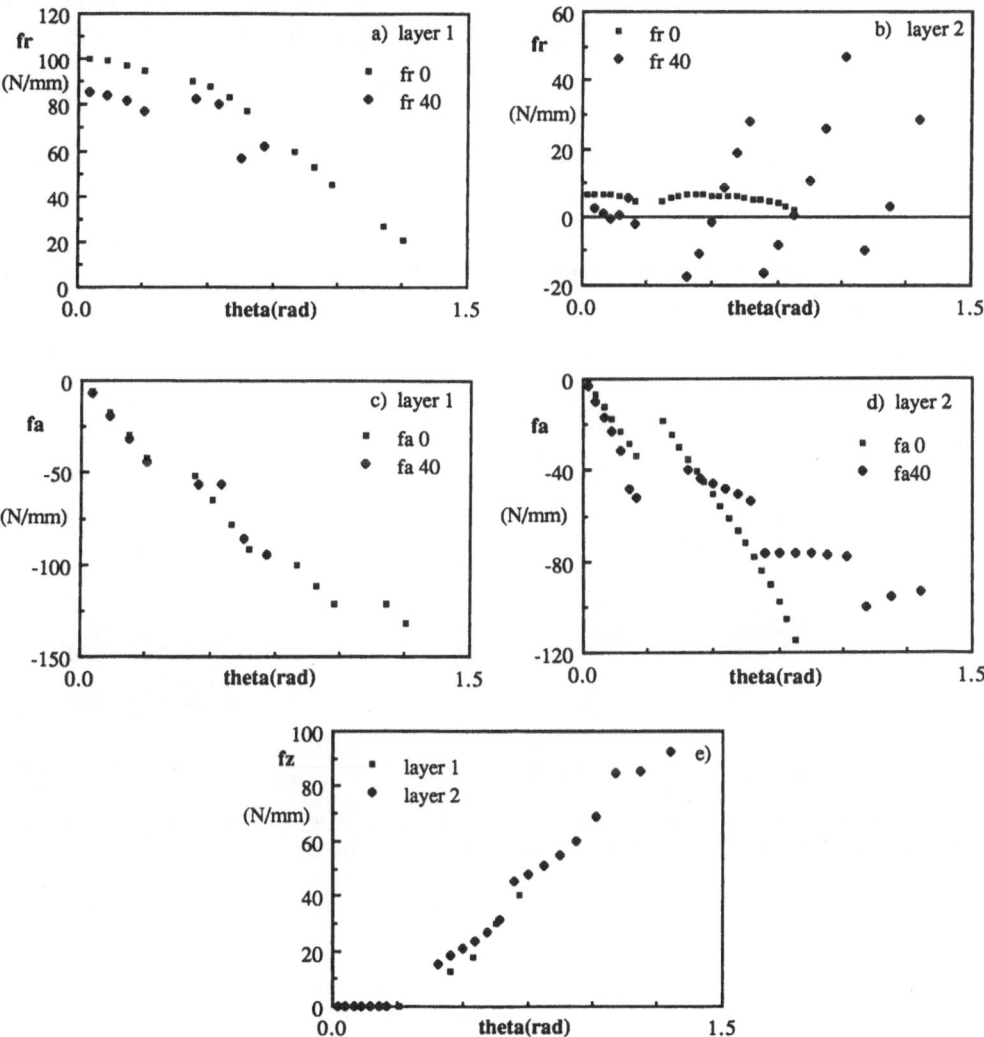

Fig.3.Radial (fr) component at z = 0 and 40 cm in a) layer 1, b) layer 2; azimuthal (fa) component at z = 0 and 40 cm in c) layer 1, d) layer 2, and e) longitudinal (fz) component of Lorentz force at z = 40 cm in layer 1 and 2

Also the radial force on e.g. the first conductor of block 1 causes a very high radial stress of about $\sigma_{1r} = 34$ MPa. These values require strong attention in view of the properties of the copper matrix of superconducting cables, in particular for Nb_3Sn conductors.

Table 2 shows the total Lorentz forces on the blocks from the sum of the forces on the conductors. The total longitudinal forces for an upper half coil head in the first layer and the second layer are outward and amount to $f_{1z} = 27.6$ kN, and $f_{2z} = 70.1$ kN, respectively, requiring strong end plates.

TABLE 2. The Lorentz force on the block of conductors. N_b is the block number (see Fig. 1).

	In median plane				Summed over coil head for $0 < \phi \le \pi/2$			
N_b	F_x	F_y	F_z	F_{tot}	F_x	F_y	F_z	F_{tot}
	N/mm	N/mm	N/mm	N/mm	kN	kN	kN	kN
1	403.7	-36.8	0.0	405.3	13.6	-0.3	6.5	15.1
2	440.9	-68.2	0.0	446.1	14.3	0.4	10.5	17.7
3	360.5	-81.0	0.0	369.5	8.1	-1.4	7.7	11.3
4	253.9	-43.0	0.0	257.6	2.9	-0.4	2.9	4.1
5	60.1	-116.6	0.0	131.2	4.5	-13.6	10.5	17.8
6	698.8	-809.8	0.0	1069.7	44.8	39.9	59.6	84.6

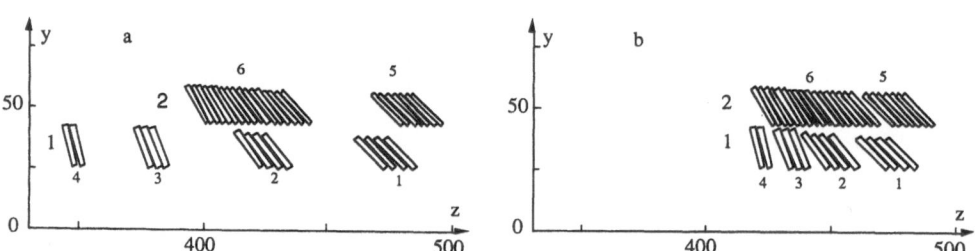

Fig.4. Layout of end of coil head for a) open and b) closed configuration; dimensions are given in mm

TABLE 3 Comparison of the sextupole component of the field integral for the open and closed configuration of Fig.4

Length	Config.	b_3 total	b_3 straight	b_3 head
1m	open	$9.17*10^{-3}$	$1.07*10^{-2}$	$-1.50*10^{-3}$
1m	closed	$-4.11*10^{-4}$	$1.09*10^{-3}$	$-1.50*10^{-3}$

For the HERA dipole magnets[6] a study was made of the influence of the spacing between conductors in the coil heads on the maximum field in the conductor locally. The contribution of the sextupole component to the field integral was minimized locally at the coil head. For the LHC magnet, the sextupole contributions of the straight part of the strands and those of the coil heads have opposite sign for the configuration of Fig.1 and a closed configuration of the blocks, as shown in Table 3. This permits the design of a coil head with a minimal sextupole component.

Table 4 shows the comparison[1] between the multipoles for a 1000m magnet using our method and a two-dimensional POISSON calculation for the configuration of Fig.1. Taking the same current for both cases, the direct contributions of the strands to $B(0,0,0) = 10T$ are the same. Scaling of the mirror contribution results in an average relative permeability $\mu_r = 16.3$ for our calculation, which is used to recalculated the coefficients for the 1000m magnet $(1000m, re)$. We observe the largest discrepancies for the sextupole and decapole coefficients which are particularly sensitive for the layout of the coil heads and are not described by the POISSON calculation. The POISSON calculation also shows that the configuration of Fig.1 has not yet resulted in sufficiently small multipole coefficients for $n \geq 3$. A similar comparison[1] with a three-dimensional TOSCA calculation yields $\mu_r = 18.4$.

TABLE 4. Comparison of the multipole expansion (1) of the Poisson calculation and a calculation for a 1000m magnet

n	b_n Pois	b_n 1000	b_n1000re	n	b_n Pois	b_n 1000	b_n1000re
1	$1.00*10^{+0}$	$1.00*10^{+0}$	$1.00*10^{+0}$				
3	$6.21*10^{-4}$	$-6.56*10^{-5}$	$-1.06*10^{-4}$	13	$-8.66*10^{-5}$	$-4.84*10^{-6}$	$4.58*10^{-6}$
5	$8.81*10^{-5}$	$-2.76*10^{-6}$	$2.11*10^{-5}$	15	$9.68*10^{-5}$	$1.77*10^{-4}$	$1.80*10^{-4}$
7	$9.39*10^{-4}$	$9.51*10^{-4}$	$9.73*10^{-4}$	17	$-2.80*10^{-4}$	$-2.37*10^{-4}$	$2.42*10^{-4}$
9	$6.71*10^{-4}$	$6.01*10^{-4}$	$6.05*10^{-4}$	19	$8.52*10^{-5}$	$-6.43*10^{-5}$	$6.63*10^{-5}$
11	$4.67*10^{-4}$	$5.53*10^{-4}$	$5.66*10^{-4}$	21	$4.03*10^{-5}$	$4.90*10^{-5}$	$5.02*10^{-5}$
I	14941A	14600A	14941A	μ_r varying	∞	16.3	

REFERENCES

1. C. Daum and D.ter Avest, Three-dimensional computation of the magnetic fields and Lorentz forces of an LHC dipole magnet, NIKHEF-H 89/12 and LHC note No. 94.
2. The large Hadron Collider in the LEP Tunnel, Eds.G.Brianti and K. Hübner, CERN 87-05, 27 May 1987.
3. D. Leroy, R.Perin, G. de Rijk, W. Thomi, Design of a High Field Twin Aperture Superconducting Dipole Model, CERN SPS/87-32 (EMA), LHC note No 62.
4. TOSCA, Vector Fields, Oxford, UK.
5. POISSON, CERN program library, T604.

6. K.H. Mess, P. Schmüser, Superconducting Accelerator Magnets, DESY HERA 89-01, and in the Proceedings of the CERN Accelerator School on Superconducticity in Particle Accelerators, Ed. S. Turner, CERN 89-04, p.87.

COOLING OF THE SYNCHROTRON RADIATION SHIELD

IN THE ELOISATRON MAGNETS

Lorenzo Resegotti

chemin Edmond Rochat 3 - CH 1217 Meyrin

INTRODUCTION

Heating by synchrotron radiation from a proton beam starts to create problems at high beam intensities in the superconducting magnets of the Superconducting Super Collider (SSC)[1] in the USA and of the Large Hadron Collider (LHC)[2] at CERN, but its consequences are conspicuous in the ELOISATRON design[3], even at the modest beam intensity foreseen in the present parameter list[4]. For a preliminary evaluation, the parameters in table 1 are assumed. With these parameters, the synchrotron radiation at top energy would have a critical energy of 10.7 keV, and the radiated power per unit length of dipole field would be 1.84 W/m. Such a heat load would be unacceptably large for the ELOISATRON magnets, which are assumed to be cooled by helium II at 1.8 K (with a planned specific load of less than 0.2 W/m). It is therefore essential for the synchrotron radiation power to be almost entirely intercepted by a shield, cooled by ordinary helium. For obvious reasons of space this shield would best consist, at least inside the magnets, of the beam chamber itself.

Table 1. Selected ELOISATRON parameters

Maximum proton energy	100 TeV
Maximum dipole field	10 T
Length of a normal half-period	100 m
Bending angle per normal half-period	2.35 mrad
Number of dipole magnets per half-period	6
Magnetic length of a dipole magnet	13.1 m
Magnetic length of a quadrupole	13.6 m
Phase advance per period	$\pi/3$
Normalized beam emittance	$0.75\pi \cdot 10$ rad·m
Circulating beam current	16.43 mA
Required aperture for the beam	14.8 mm

We assume therefore the beam chamber to be thick enough to absorb almost entirely the synchrotron radiation power: its cooling system will have also to absorb the environmental heat intake from its supports and connections and the power dissipated in the chamber wall by beam induced currents and by protons lost from the beam. Considering the heat load of the beam chamber in a half-period of the normal machine lattice, 100 m long, we remark that heat will come to it from the environment only at the junctions between magnets and in the short straight section space used by service connections, which add up in total to a length of less than 10 m. By comparison with the SSC and the LHC designs, the environmental heat load can therefore be estimated to be less than 4 W. The beam chamber will be internally coated with copper, as in the SSC, so that the heat dissipation by beam induced currents can be evaluated, by comparison with SSC values, to be less than 0.1 W over the half-period. Beam losses due to beam scattering by the residual gas will probably be less than 0.2 W, but the chamber cooling system will have to cope also with the effects of other proton losses, as long as they do not produce resistive transitions (quenches) in the magnets. On the basis of the results of the computations made for the LHC shielding[5], we assume the corresponding power dissipation in the chamber wall over a half-period to be of the order of 10 W. All these heat loads are small with respect to the synchrotron radiation from the dipoles, which is about 145 W (in fact, a little less, because of the field reductions at the ends). The contribution from a quadrupole will not exceed 0.5 W, under normal closed orbit conditions. The chamber cooling system can therefore be designed for a maximum power dissipation of 160 W over a half-period.

At maximum beam energy, the synchrotron radiation hits the beam chamber mostly on a strip less than 0.5 mm high, because the beam height ($4\sigma_z$) at 100 TeV varies between 0.06 mm and 0.1 mm in the normal lattice and the angular spread of the radiation in the vertical plane ($4\ \sigma_\psi$), which is about 0.016 mrad, gives a height of 0.40 mm at the point of impact, about 25 m downstream. This heat concentration would create difficulties with a beam chamber made out only of stainless steel, which has a very poor conductivity at liquid helium temperature (about 0.3 $W \cdot m^{-1} \cdot K^{-1}$). An efficient cooling and an acceptable temperature distribution in the beam chamber are made possible by the inner copper coating, which is also needed to reduce the power losses due to beam induced image currents: its thermal conductivity at 4.5 K is about 1000 times that of stainless steel.

Because of the large amount of synchrotron radiation power, a copper coating 0.2 mm thick is desirable in the ELOISATRON: by applying it inside a stainless steel pipe of 16 mm inner diameter, a free diameter of 15.6 mm is obtained, which is adequate to ensure the required aperture of 14.8 mm for the beam, taking chamber misalignments and dimensional tolerances into account. A 2 mm wall thickness, which is desirable to absorb a large enough fraction of the synchrotron radiation power, is adequate to enable the stainless steel pipe to withstand the bursting forces exerted by the magnetic field on the eddy currents induced into the 0.2 mm copper layer by the rapid field drop in case of magnet quench. With these dimensions, it can be anticipated that the temperature variation along the copper coating will not exceed 0.2 K. The temperature difference between the copper coating and the cooling fluid depends on the layout of the cooling system, but it is expected to remain lower than 1 K for all solutions considered in this paper.

COOLING BY HELIUM VAPORIZATION

For the purpose of a preliminary evaluation of the beam chamber cooling system, which will have eventually to be accurately streamlined because it takes a part of the very valuable space inside the magnet aperture, we assume it to consist of stainless steel pipes, welded on the periphery of the beam chamber following a slightly helical pattern, in order to ensure a uniform sharing of the heat input from the synchrotron radiation. We assume liquid helium at 200 kPa (2 bar) and 4.5 K to be supplied to the pipes at each straight section and gaseous (or two-phase) helium to be removed at the following straight section, that is at 100 m distance. The helium gas return line is supposed to be at 4.5 K and 120 kPa (1.2 bar).

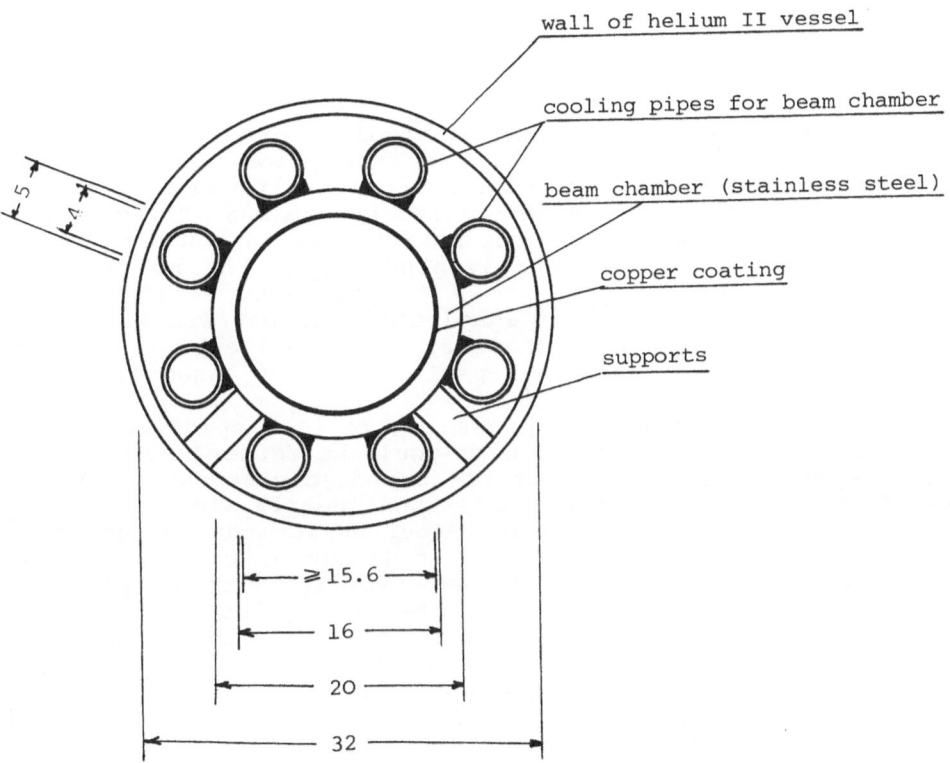

Fig. 1. Cross section of beam chamber with welded cooling pipes (helical pattern).

The specific enthalpy[6] of the liquid at 4.5 K, 200 kPa is 11.21 kJ/kg. If the final state of the fluid, at the end of the half-period, is saturated vapour at 4.5 K, with p = 130 kPa and specific enthalpy 29.81 kJ/kg, the increase of specific enthalpy is 18.6 kJ/kg. As the specific kinetic energy of the fluid remains negligible with respect to this value (less than 1%), the whole amount can be considered available for cooling purposes. The required helium mass flow rate is therefore: $\dot{m}$ = 160/18600 kg/s = 8.60 g/s.

In order to evaluate the required size of the cooling pipes, the pressure gradients at fluid input and output corresponding to the required mass flow rate have been computed initially for the simple case of n pipes of circular cross section, of inner diameter D, so arranged in parallel as to share uniformly the heat input and the helium flow, trying different values of n and D. Table 2 shows, for a helium mass flow rate of 8.6 g/s, the main steps and the results of these preliminary computations. The friction coefficient f in the formula of pressure gradient[7]

$$\frac{dp}{ds} = f \frac{V^2}{v\ D} = \left(\frac{4}{\pi}\right)^2 f \frac{\dot{m}^2\ v}{n^2\ D^5}$$

(where V is the velocity of the fluid, v its specific volume, and SI units are used throughout), is a function of Reynolds number; the values used are based on Nikuradse's results for commercial pipes of relative roughness b/D, as reported in reference 8. In our example, an average wall roughness b = 1.5 μm has been assumed.

The total pressure drop in the cooling pipes over a half-period could be computed under the assumption that the two-phase fluid remains a homogeneous mixture of liquid and vapour, the properties of which are determined by the mass ratio of the two phases. The effect of junctions and bellows could be taken into account by an equivalent lengthening of the normal pipe which, in our case, amounts to a few percent. Another small lengthening comes from the slightly helical pattern which the pipes follow along the beam chamber. An overall equivalent pipe length of 110 m over 100 m half-period has been assumed.

The integration of the pressure gradient dp/ds, as shown in the above formula, along a pipe would be rather tedious, because the specific volume v of the supposedly homogeneous mixture of liquid and vapour increases nonlinearly along the pipe, as a result of the simultaneous vapour enrichment and decrease of pressure; also the coefficient f is not constant. For our preliminary evaluation, we have arbitrarily used the products of the equivalent pipe length by the average of the pressure gradients computed for the pure liquid and vapour at input and output respectively, as shown in the last line of table 2. Inspection of these figures already shows that a pipe diameter of 3 mm would be too small and that at least 8 pipes with a diameter of 4 mm (or 6 pipes of 4.5 mm, or 4 pipes of 5 mm) would be necessary to allow for the required mass flow rate of the cooling fluid under the available pressure drop of 70 kPa.

Table 2. Limit pressure gradients for cooling with two-phase helium in pipes of different numbers and diameters

---

Helium mass flow rate $\dot{m}$ = 8.60 g/s

Properties of fluid		at input	at output
Phase		liquid	vapour
Temperature	T (K)	4.5	4.5
Pressure	p (kPa)	200	130
Specific volume	v (m^3/kg)	$8.04 \cdot 10\ 3$	$45.2 \cdot 10$
Specific enthalpy	h (kJ/kg)	11.21	29.81
Viscosity	η (Pa·s)	$3.24 \cdot 10\ 6$	$1.38 \cdot 10$

Number of pipes	n	4	6	6	8	12
Pipe diameter	D (m)	0.005	0.0045	0.004	0.004	0.003

Input Conditions

Velocity	V (m/s)	0.88	0.72	0.92	0.69	0.82
Reynolds number	Re	169000	125000	141000	106000	96000
Friction coeff.	f	0.0090	0.0094	0.0093	0.0097	0.0102
Pressure gradient	$\frac{dp}{ds}\left(\frac{kpa}{m}\right)$	0.17	0.136	0.24	0.14	0.28

Output conditions

Velocity	V (m/s)	4.95	4.07	5.16	3.87	4.58
Reynolds number	Re	396000	293000	330000	248000	220000
Friction coeff.	f	0.0082	0.0085	0.0086	0.0088	0.0093
Pressure gradient	$\frac{dp}{ds}\left(\frac{kpa}{m}\right)$	0.89	0.69	1.27	0.73	1.44

$$\Delta_p \approx \frac{110}{2}\left\{\left(\frac{dp}{ds}\right)_i + \left(\frac{dp}{ds}\right)_0\right\} (kPa)$$

	58.3	45.4	83.0	47.8	94.6

A more accurate computation with the homogeneous fluid hypothesis would lead to somewhat lower values of the total pressure drop, but would not be really useful, because the hypothesis is known experimentally to be too optimistic and all deviations from the ideal flow conditions lead to substantially larger pressure drops. In general practice, ad-hoc safety factors are used[7]: this approach would confirm the conclusions of the preliminary inspection.

The number of pipes which can be fitted around the beam chamber is limited by the required access for continuous welding, which is desirable for good thermal contact: the helical disposition must also be accounted for. Assuming 8 pipes of 4 mm inner diameter, o.5 mm thick, on a beam chamber of 20 mm outer diameter, as shown in figure 1, the whole system would occupy a cylindrical space of 30 mm diameter. With a clearance of 1 mm to the wall of the helium II vessel, the diameter of the latter would be 32 mm.

COOLING BY SUPERCRITICAL HELIUM

A better thermodynamic efficiency in the removal of the synchrotron radiation power can be achieved by the use of supercritical helium as the coolant, with the additional advantage of single-phase flow. The example presented in table 3 has been taylored to the same system of 8 pipes of 4 mm inner diameter as shown in figure 1. It corresponds to an input temperature of 8 K and an input pressure of 500 kPa, with an available pressure drop of 100 kPa over the 100 m long half-period. Allowing for an output temperature of 9.8 K, the available specific enthalpy increase would be 15 kJ/kg, so that a mass flow rate m = 10.67 g/s would be needed. Table 3 shows the results of the computation of the pressure gradients at input and output. The total pressure drop can be expected to be slightly lower than the value of 98 kPa, based on their arithmetical average: a computation in ten steps of equal enthalpy increase gives actually 94 kPa. Therefore the available pressure drop of 100 kPa can be considered adequate, as the single-phase flow conditions do not require large safety margins.

Table 3. Limit pressure gradients for cooling with supercritical helium in a set of pipes welded to the beam chamber

Number of pipes	$n = 8$
Inner diameter of pipes	$D = 4$ mm
Wall roughness	$b = 1.5$ μm
Helium mass flow rate	$m = 10.67$ g/s

		at input	at output
Temperature	T (K)	8	9.8
Pressure	p (kPa)	500	400
Specific volume	v (m^3/kg)	$24.0 \cdot 10^{-3}$	$44.7 \cdot 10^3$
Specific enthalpy	h (kJ/kg)	43.50	58.50
Viscosity	η (Pa·s)	$2.38 \cdot 10^{-6}$	$2.46 \cdot 10^6$
Velocity	V (m/s)	2.55	4.74
Reynolds number	Re	178000	172000
Friction coefficient	f	0.0092	0.0092
Pressure gradient	$\dfrac{dp}{ds}\left(\dfrac{kPa}{m}\right)$	0.62	1.16

$$\Delta p = 110 \cdot \frac{0.62 + 1.16}{2} \text{ kPa} = 97.9 \text{ kPa}$$

An alternative solution, which would allow to reduce a little the space requirements, would consist in surrounding the beam chamber by a coaxial pipe, supported by radial fins, which could be formed outside the chamber by means of a special drawing die. Figure 2 shows an example with 3 radial fins. The outer pipe would fit loosely onto the fins, to allow for easy assembly. In order to produce a good mixing of the fluid all around the pipe, substantial stretches of the fins would be machined off along a helical pattern and, if necessary, small slanted deflectors could also be welded at some places. Despite these precautions, the two-phase fluid in such a duct of annular cross section with non-uniform heat input along the internal periphery would probably be inhomogeneous and the risk of serious perturbations of the flow due to instabilities would not be negligible. The use of a single-phase coolant would be much preferable: the dimensions of figure 2 have been found suitable for supercritical helium under the same limit conditions of temperature and pressure as adopted in the example of table 3.

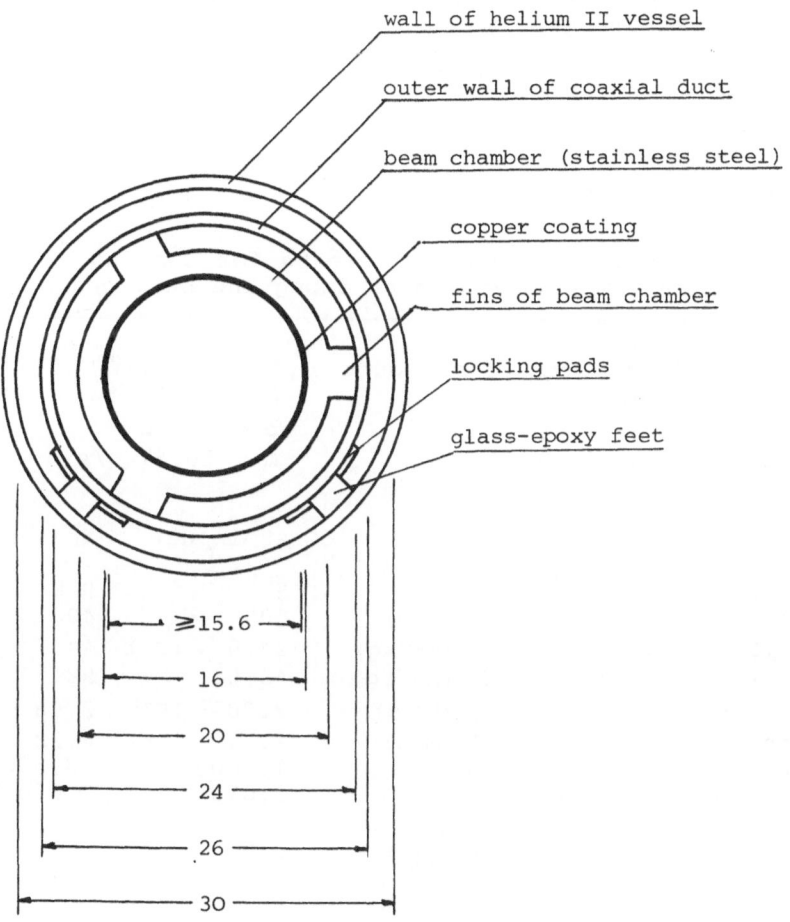

Fig. 2. Cross section of beam chamber with coaxial cooling duct for supercritical helium (8 K input).

In the case of cooling ducts of non-circular cross section, the frictional resistance to homogeneous fluid flow can be evaluated in terms of the hydraulic diameter D_h, which is defined as four times the ratio of the cross-sectional area S to the perimeter P of the cross section: $D_h = 4S/P$. The pressure gradient can be computed by means of the formula:

$$\frac{dp}{ds} = k \, f \, \frac{V^2}{v \, D_h}$$

where f is still given, as a function of the Reynolds number and of the relative wall roughness b/D_h, by Nikuradse's results, as reported in reference 8. The coefficient k, for ducts of annular cross section, is a slowly-varying function of the Reynolds number, and for all flow conditions in this example can be assumed[8] to be k = 1.06. For the dimensions in figure 2 and taking the 3 fins into account as if they were continuous, it is $S = 133$ mm^2, $P = 125$ mm, $D_h = 3.62$ mm.

The results of the computation of pressure gradients at input and output are given in table 4. As in the previous case, the total pressure drop will be slightly lower than the value of 93 kPa, computed using their arithmetical average: the result confirms the good match between the chosen dimensions of the duct and the assumed input and output conditions of the coolant.

Table 4. Limit pressure gradients for cooling with supercritical helium in a coaxial duct along the beam chamber

Cross-sectional area of the duct		$S = 113$ mm^2	
Hydraulic diameter		$D_h = 3.62$ mm	
Wall roughness		$b = 1.5$ µm	
Helium mass flow rate		$\dot{m} = 10.67$ g/s	

		at input	at output
Temperature	T (K)	8	9.8
Pressure	p (kPa)	500	400
Specific volume	v (m^3/kg)	$24.0 \cdot 10^{-3}$	$44.7 \cdot 10^{-3}$
Specific enthalpy	h (kJ/kg)	43.50	58.50
Viscosity	η (Pa·s)	$2.38 \cdot 10^{-6}$	$2.46 \cdot 10^{-6}$
Velocity	V (m/s)	2.27	4.22
Reynolds number	Re	144000	139000
Friction coefficient	k-f	0.010	0.010
Pressure gradient	$\frac{dp}{ds} \left(\frac{kpa}{m}\right)$	0.59	1.10

$$\Delta p = 110 \cdot \frac{0.59 + 1.10}{2} \text{ kPa} = 93 \text{ kPa}$$

The outer diameter of the external wall of the coaxial duct in figure 2 is 26 mm, to be compared with the 30 mm outer diameter of the cylindrical space occupied by the multipipe system in figure 1. However, the gain in the diameter of the helium II vessel is only 2 mm, because the beam chamber complex is not rigid enough to be supported only at the ends of the magnets. The intermediate supports, which rest on the inner pipe of the helium II vessel, are longer if they fit between the cooling pipes than if they stem from the outer wall of the coaxial duct. In the latter case a clearance of 2mm is needed to keep the heat leak by conduction below 10 mW per meter of duct, even using glass-epoxy feet which have a very low thermal conductivity up to 10 K.

It might be feared that, with a beam pipe temperature in the vicinity of 10 K, the cryopumping efficiency for hydrogen desorbed by synchrotron radiation would be inadequate. Fortunately, experience gained at CERN with cooling of vacuum chambers, thoroughly treated for ultrahigh vacuum but without final bakeout in situ, to cryogenic temperatures has shown surface coverages by adsorbed hydrogen equal to small fractions of a monolayer[9]. At coverages below one monolayer, the H_2 condensate is strongly adsorbed to the substrate and the observed pressure, which is generated by incident radiation, is much lower than the saturated vapour pressure[10]. This situation is expected to persist, practically independent of the substrate temperature, up to about 10 K[9]. Obviously, confirmation of these expectations should be sought by ad-hoc experiments with radiation of the appropriate intensity and spectral composition, but the possibility of cooling with supercrital helium between 8 and 10 K in a coaxial duct is meanwhile worth considering, because of its simplicity and smaller space requirements.

Should experimental results show that a better cryopumping efficiency is necessary, supercritical helium would have to be used in the multipipe arrangement, with holes in the beam chamber wall between pipes, to make the beam region communicate all along with the external vacuum region, limited by the inner wall of the helium II vessel, which is at 1.8 K.

As the power dissipation by synchrotron radiation in the ELOISATRON will vary from a negligible value at injection energy to a high value at maximum operating energy (and will be zero, as all beam related power losses, during the intervals between cycles) an important effect for the operation of the cooling system is the variation of the mass flow rate during the machine cycle. The thermodynamic transformation associated with the increase of the fluid enthalpy modifies the resistance to flow, so that the mass flow rate under a constant pressure drop decreases as the heat input increases. This characteristic is obviously the contrary of what one would like: the worst situation, in this respect, is met with cooling by helium vaporization, where the mass flow rate may decrease by almost a factor 2 in going from no load to full load.

For the case of cooling by supercritical helium in a coaxial duct, exemplifield in table 4, the diagrams of figure 3 show helium mass flow rate and output temperature as functions of total input power over a half-period, for an input

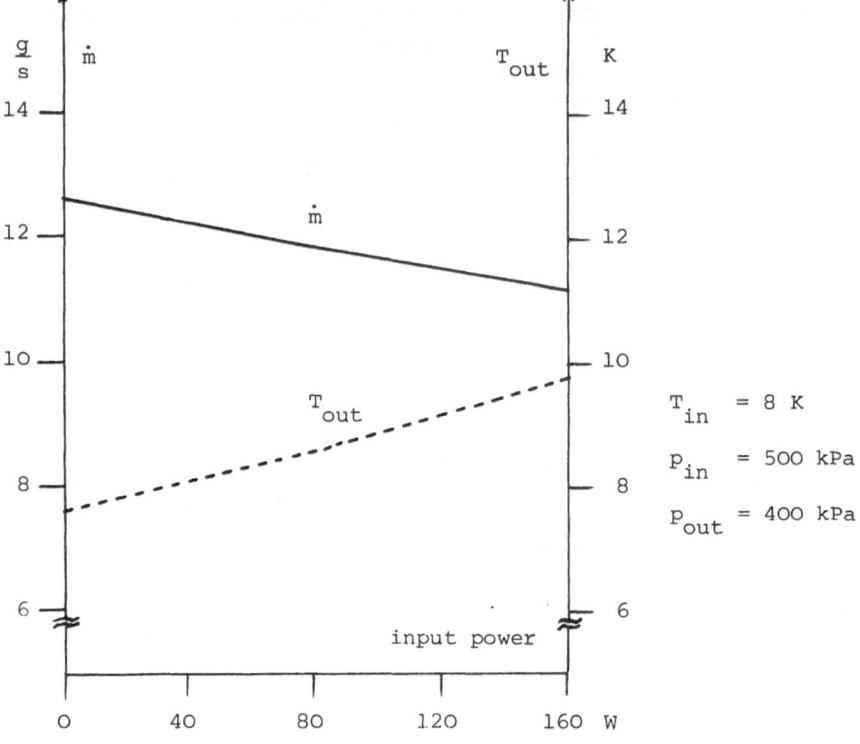

Fig.3. Mass flow rate and output temperature of super-critical helium, in the cooling duct of figure 2, as a function of the input power over a half-period (100 m).

temperature of 8 K and under a constant pressure drop from 500 kPa to 400 kPa. (It is worth noting that, at zero load, the helium would cool to about 7.6 K in its isoenthalpic expansion.) The mass flow rate at full load is 12% lower than at zero load. The diagrams for supercritical helium in a multipipe system would be very similar. As these cooling systems require a cold pump and a heat exchanger at each half-period, the operating conditions will depend also on the characteristics of the cold pump, but figure 3 provides an indication that the situation ought to be manageable.

Further studies and experimental tests may reveal other drawbacks of the cooling system using supercritical helium at 8 K, but at the present stage its better thermodynamic efficiency appears still to recommend it for serious consideration, leaving open the choice of either the coaxial duct (figure 2 and table 4) or the multiple-pipe (figure 1 and table 3) solution.

Additional remark

As pointed out by R. Meinke during the discussion of this paper, the helium pressure in the beam chamber cooling system will considerably increase at magnet quench, due to energy dissipation by the eddy currents induced in the 0.2 mm copper coating by the fast decay of the magnetic field. The pipes are mechanically quite strong because of their small ratio of diameter to thickness, and automatic gas exhausts can be provided at their ends, but, should the pressure rise be nevertheless a matter of concern, the dissipation could be reduced by reducing the copper thickness. Even a reduction to as little as 0.05 mm would not have intolerable consequences for the transmission of the heat deposited by the synchrotron radiation. In the case of the coaxial duct, for instance, the temperature difference across the stainless steel wall would remain everywhere lower than 1 K, although it would vary azimuthally by more than a factor 5. As for the power dissipation by beam induced currents, which in this case would be about a factor 4 larger, its value would still be small with respect to the synchrotron radiation power at maximum beam energy.

REFERENCES

1. SSC Central Design Group, Conceptual design of the Superconducting Super Collider, SSC-SR-2020 (March 1986).
2. G. Brianti and K. Hubner, ed., The Large Hadron Collider in the LEP tunnel, CERN 87-05 (May 1987).
3. A. Zichichi, project leader, The ELOISATRON project (June 1988).
4. K. Johnsen, The ELOISATRON, in this book.
5. L. Burnod and J. Gareyte, LHC parameters, CERN/SPS/Int. Note EMA 89-7, LHC Note 92 (May 1989).
6. R.D. McCarty, Thermophysical properties of helium-4 from 2 to 1500 K with pressures to 1000 atmospheres, NBS Technical Note 631 (1972).
7. B.A. Hands, Fluid dynamics, in."Cryogenic Engineering", B.A. Hands, ed., Academic Press, London (1986).
8. I.E. Idel'chik, "Handbook of Hydraulic Resistance", Gosundarstvennoe Energeticheskoe Izdatel'stvo, Moskva (1960), Israel Programme for Scientific Translations, Jerusalem (1966).
9. C. Benvenuti (CERN), Private communication.
10. C. Benvenuti, R.S. Calder and G. Passardi, The influence of thermal radiation on the vapour pressure of condensed hydrogen between 2 and 4.5 K, CERN divisional report ISR-VA/76-19 (1976).

PROBLEMS ARISING FROM BEAM LOSSES

IN SUPERCONDUCTING COLLIDERS

L. Burnod

CERN, European Organization for Nuclear Research

Geneva, Switzerland

INTRODUCTION

Beam losses are a very important problem for superconducting (SC) high-energy accelerators and colliders. When protons hit the vacuum chamber, cascades of particles are produced and radiated energy is deposited in the surroundings materials, in particular in the SC coils. This energy can heat the coil above the critical temperature, causing it to go normal and the magnet to quench.

Several processes contribute to beam losses (Chapter 1). Numerical values of the energy deposit per incident proton are given (Chapter 2). The thermal behaviour of a coil is analysed (Chapter 3). In steady state the result is a temperature gradient which should not exceed a certain value corresponding to the quench limit (Chapter 4).

To protect the SC magnets, set of collimators could be envisaged. But they are not perfect absorbers and secondary particles escape, mainly due to the edge effect. The collimator efficiency is strongly related to how the beam emittance grows up (Chapter 5).

1. ORIGIN OF LOSSES

Some losses can be considered as purely accidental and produce transient phenomena, while some others systematically occur in steady state operation.

1.1 At injection, if the initial conditions have changed, the full beam could hit the vacuum chamber during the first turn. The loss would occur at an unknown location of the machine, determined by the residual perturbations, and in a very short time, i.e one bunch of a few nanoseconds.

The duration of the loss being small compared to the time constant of any heat exchange with the surroundings, it can be considered as a pure adiabatic process, where the copper enthalpy reserve is the dominant factor (see Ch.3) and determines the maximum tolerable number of lost protons, of the order of one single bunch[1]. Checking the status of the machine with only one single bunch before injecting the full intensity seems a wise precaution. Even if this bunch is lost, the next one is coming when a new injection is requested, one can assume that there is no cumulative effect.

1.2 During the accelerating ramp or at top energy in collider mode, accidental beam losses can occur when crossing a resonance or due to the misfunctioning of a magnetic element or a bad RF manipulation. With the experience of existing accelerators, the corresponding losses occur at a specific location, which for any reason is an aperture restriction, much more likely than a distributed loss all around the machine.

    All the magnetic elements being superconducting including the correcting dipoles, their ramping time are greater than 1 mn. This time scale is then of the same order or slower than the expected time constants of the heat diffusivity inside the cables and between cables and circulating helium through the insulation. Transient phenomena have to be considered, but up to now they are not well under control.

1.3 <u>In collider mode, systematic losses will also occur</u>

    Due to beam-gas interactions, distributed losses are expected all around the accelerator, but almost all of the vacuum chamber is at liquid helium temperature. It is so a large cryopump reducing drastically the pressure and the effect of scattering on the residual gas can be neglected, The example of the LHC is given in Annex I.

    Due to beam-beam collisions, important losses are produced in the intersection regions of high luminosity. A large fraction is dissipated in the detector and in the low-ß insertion magnets, while a non negligible part of the collisions participates to a blow-up of the beam emittance and then contributes to losses elsewhere along the arcs.

    Elastic or quasi-elastic scattered protons, produced in hadron collisions, are characterized by small energy losses and small scattering angles, which permits them to travel with the beam for considerable long distances, so a significant fraction is lost eventually at one or more aperture limiting restrictions. These losses are then strongly sensitive to geometric factors such as magnet alignment and closed orbit position.

    Considering only the hadron collisions, the p-p cross-sections (total and elastics) are given in Table I for the SPS and the LHC[2]. In the interaction regions (I.R) where there is no physics experiments, the hadron collisions are neglected, the two beams being vertically separated or the I.R being set to a modest low-beta configuration. Only one LHC physics experiment is assumed to work at a given time. The SPS intensity corresponds to average p and pbar values obtained during the 1989 p-pbar runs. LHCN corresponds to the nominal LHC intensity and LHCM corresponds to the maximum expected luminosity. The beam lifetime due to collisions, $Tc = I (dt/dI)$, is calculated by assuming that all the particles having made a nuclear collision are lost.

Table 1

	SPS	LHC	LHCM
Momentum (Gev/c)	315	8000	8000
Total cross-section (σ_{TOT}) (mb)	63	140	140
Elastic cross-section (σ_{ELA})(mb)	14	40	40
a (mb/(Gev/c)^2)	210	805	805
$1/b = (p* \theta^\circ)^2$ (Gev/c)^2	0.0667	0.0435	0.0435
θ° (mrad)	0.82	0.026	0.026
Nr of interaction regions	2	1	1
Nr of bunches / beam	6	3564	4875
Nr of p / bunch ($*10^{11}$)	1.6	0.26	1.0
Np / beam	1×10^{12}	0.91×10^{14}	4.87×10^{14}
L / beam (cm$^{-2}*$s^{-1})	2.1×10^{30}	0.14×10^{34}	3.9×10^{34}
Nuclear events / s	2.6×10^5	2.0×10^8	5.5×10^9
Elastic events / s	5.8×10^4	5.6×10^7	1.6×10^9
Power in nuclear events (W)	0.026	510	14000
Power in elastic events (W)	0.006	146	4000
Beam lifetime (hrs)	1070	126	25
ß* in insertion (m)	0.5	1	0.25
Normalized emittance (π.mm.mrad)	10	5	15
2σ at ß* (μm)	122	24	21
2σ' at ß* (μrad)	244	24	84
Half aperture in low-ß quad (mm)	83	20	20
ß$_{MAX}$ in low-ß quad (m)	2500	1156	5019
2σ at ß$_{MAX}$ in low-ß quad (mm)	8.6	0.82	3.0
Y$_{MAX}$ in low-ß quad (mm)	29	0.88	0.92
ß$_{MAX}$ (in the arcs) (m)	107	169.5	169.5
2σ at ß$_{MAX}$ (in the arcs) (mm)	1.8	0.32	0.55
Y$_{MAX}$ (in the arcs) (mm)	6.0	0.34	0.17
% of elastics out of ±2σ'		8	<1

In the SPS, the calculated beam lifetime due to beam-beam and beam-gas collisions is about one order of magnitude greater than the measured beam lifetime which was around 50 hrs during the last p-pbar period. This suggests that other effects like beam-beam or intrabeam scattering contribute to reduce the beam lifetime, but this is still to be understood quantitatively. Also, the effects may not simply add. For example, the beam-beam effect is stronger for particles of large emittance (> 2 σ) and elastic scattering is populating the phase space mostly in the region between 0-5 σ.

2. ENERGY DEPOSIT PER INCIDENT PROTON

2.1 Angle of incidence

The energy deposit inside a material is strongly dependent on the incidence angle of the particle.

If the loss is due to a Closed Orbit perturbation during a run, this perturbation brought out of a superconducting element will grow progressively. Assuming a perturbation located near a focusing quadrupole, QF, the trajectory will hit the vacuum chamber aperture, Y_a, with an incidence angle $Y'_a = 2Y_a (1-\sqrt{\beta d/\beta f}) \sin (\mu/2))/L$, where βd, βf are the betatronic amplitudes in QD, QF, μ (=90^0) the phase advance per cell, and L the cell length.

For LHC, with L = 100 m and Y_a = 17 mm, Y'_a = 0. 24 mrad.

If the loss is due to an emittance growth, in a linear machine the ellipse of emittance is tangent to the aperture limit, Y_a, with an angle

$$Y'_a = - (\alpha/\beta) * Y_a,$$

where β is the betatronic amplitude and $\alpha = -1/2$ (dβ/ds).

The ellipse has its maximum amplitude in the center of a QF, where α = 0 = Y'_a. On both sides of QF, where the probability of hitting the aperture is high, the incidence angle can be computed.

For LHC with an aperture of 17 mm, Y'_a = 0. 34 mrad.

In high energy colliders, the incidence angle of the particle is then very small. An important part of the energy will be deposited inside the vacuum chamber and in the first layer of the SC coils.

2.2 Computation of the energy deposit

The energy deposit in a SC dipole due to one proton striking the vacuum chamber can be simulated by several programs, like CASIM[3] or FLUKA[4].

Examples of a radial and longitudinal map of the energy deposit are given on Fig.1 to 3.

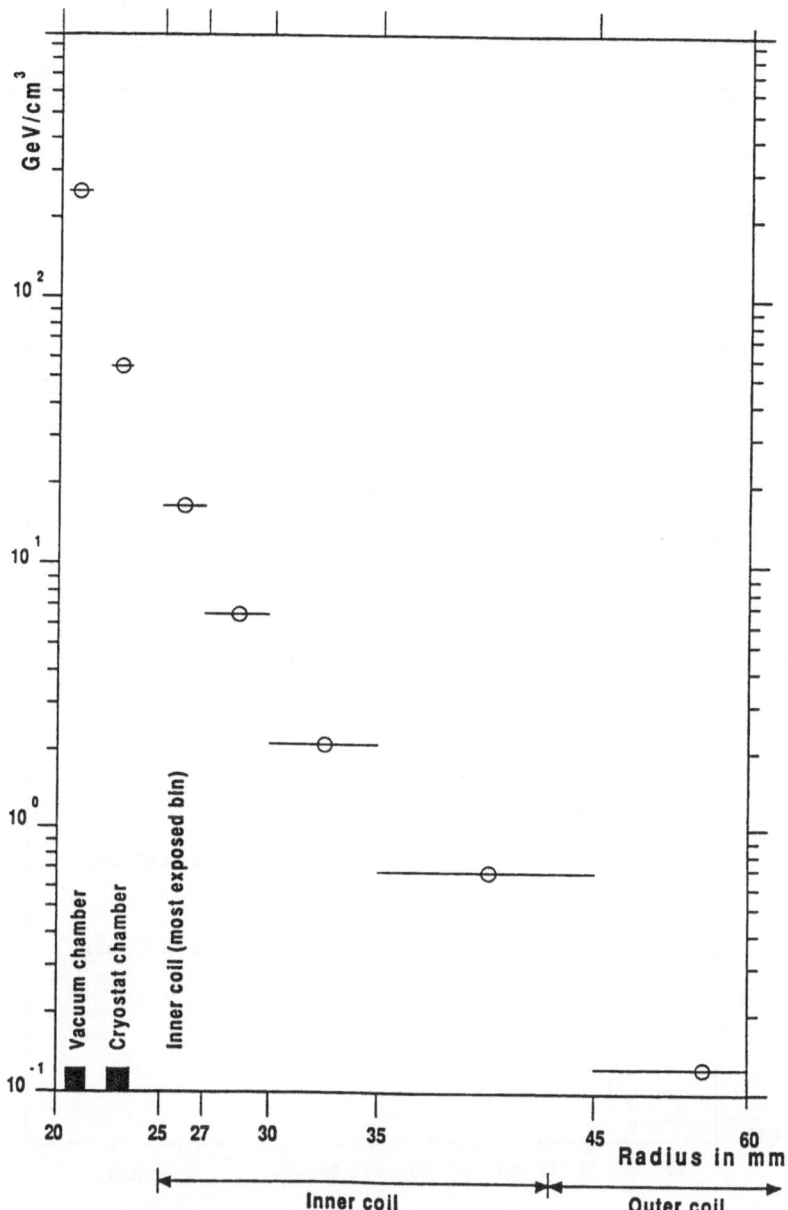

Fig.1 Radial maximum energy deposit per incident proton of 8
      Tev in a LHC magnet (Simulation with CASIM[3])
      Initial impact X = 2.05 cm   X' = 0.24 mrad   Y = Z = 0
                    0 < φ (rad) < 0.05

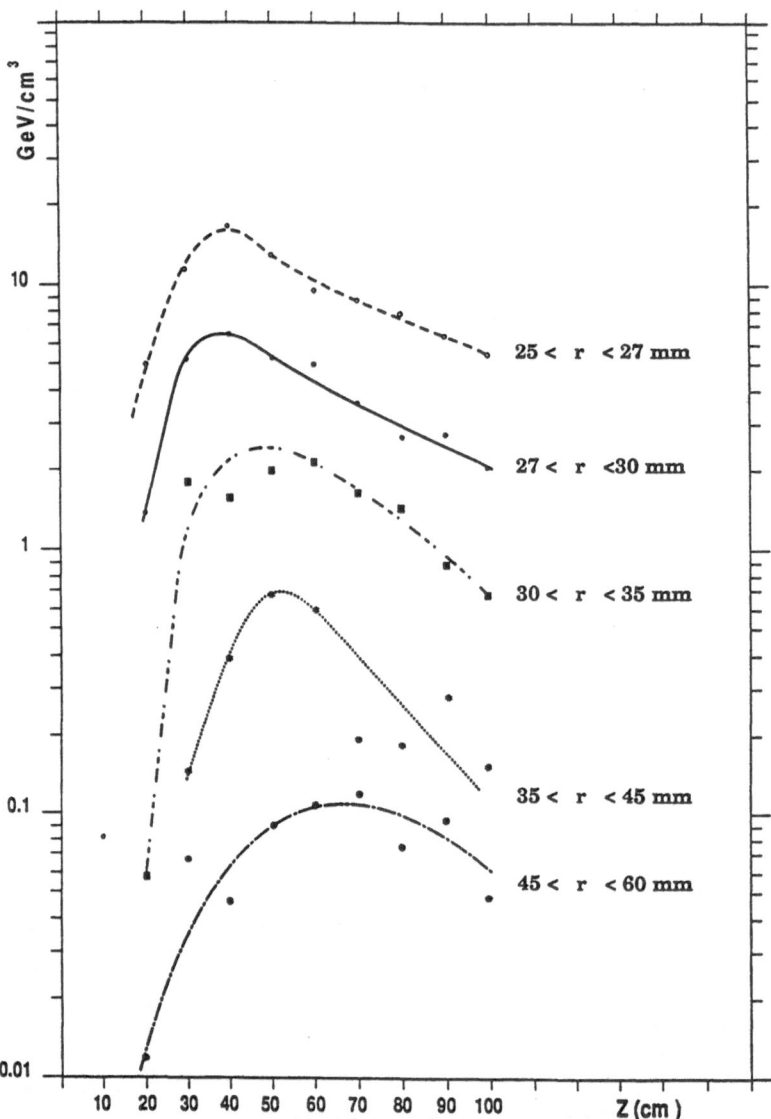

Fig.2 Longitudinal maximum energy deposit per incident proton
of 8 Tev in a LHC magnet (Simulation with CASIM[3])
Initial impact X = 2.05 cm X' = 0.24 mrad Y = Z = 0
0 < φ (rad) < 0.05

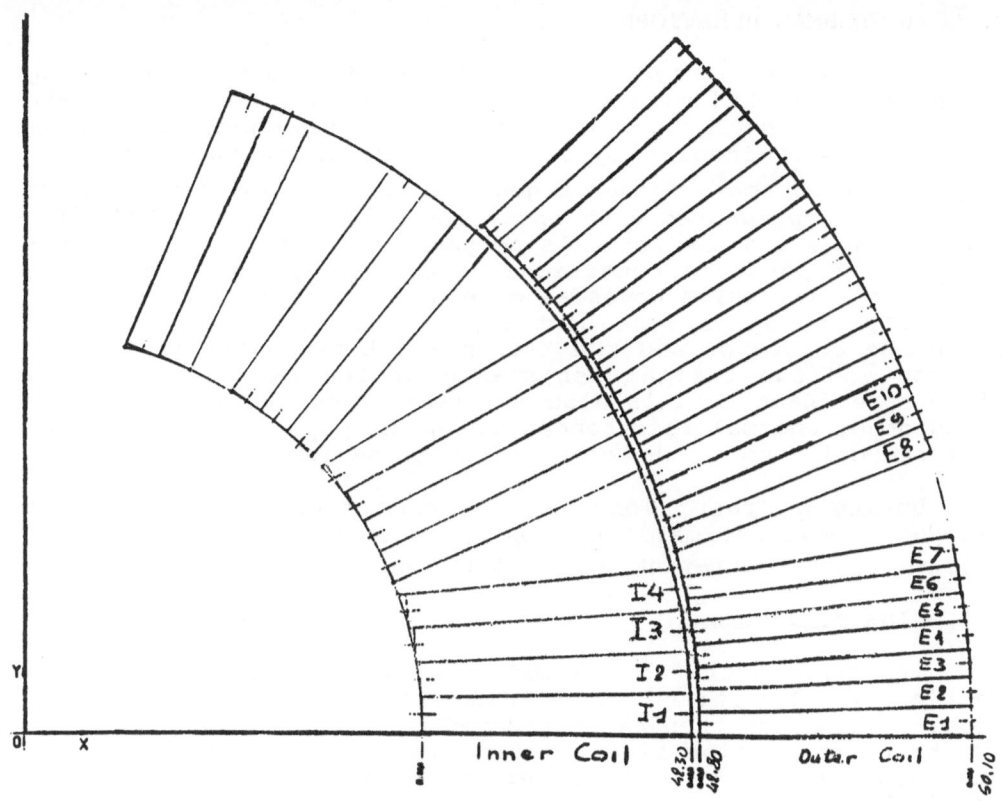

Fig.3 Energy deposit in LHC cables

Incident proton: X = 2.05 cm   Y = Z = 0
                       X' = 0.24 mrad

p = maximum energy density in the most exposed cables (Gev/cm³
    and J/cm³)

W = maximum power deposited in the most exposed cables by a
    pencil beam of $2*10^7$ incident protons per second (mW)

Cable	p (Gev/cm³)	p (10^{-10} J/cm³)	W (mW/cm³)
I-1	3.0	4.8	9.6
I-2	2.8	4.5	9.0
I-3	2.5	4.0	8.0
I-4	1.3	2.1	4.2
E-1	0.067	0.107	0.21
E-2 & E-3	0.098	0.157	0.31
E-4 to E-6	0.164	0.262	0.52
E-7 to E-10	0.124	0.198	0.40

3. COIL THERMAL BEHAVIOUR

A SC cable is made of twisted strands, each strand being a composite of NbTi filaments in a copper matrix. For the sake of electrical insulation, each cable is usually wrapped in several layers of insulating tape, for instance, two layers of 25 μm thick Kapton*. To get a good mechanical rigidity, a glass fibre tape impregnated with epoxy resin is wound round the insulated cable and the resin is polymerized under adequate pressure and temperature conditions. The consequence is a severe thermal barrier which comes as a result of the low thermal conductivity of Kapton, the high thermal contact resistance between the conductor and the Kapton and between the two Kapton layers, and the thermal resistance of the impregnated glass fibre tape. Each coil is also insulated from the collars, creating other thermal resistances along the path for heat removal from the coils up to the cold source of helium.

3.1 <u>During a "fast" loss</u>, equilibrium temperatures are not reached, local thermal time constants intervene, the enthalpy reserve of the different parts of the coils are predominant[1].

3.1.1 <u>Inside a strand</u>

The temperature is assumed to be uniform over the cross-section of the strand. The energy deposit inside a strand cross-section can then be considered as uniform. Longitudinally the heat propagation time constant is of the order of 100 μs/cm, which is too large to intervene during an "instantaneous" loss (60 ns).

3.1.2 <u>Between strands inside a cable</u>

A compromise in the cable design has to be found between mechanical stability, electrical insulation, thermal conductivity, reduction of losses due to eddy current and improved cooling by a better penetration of superfluid He into the cable (porous cable).

In an extreme case, a perfect thermal conductivity between strands could be attained by soldering the strands together. The soldering being assimilated to a copper bridge of the same diameter as the strand, the thermal equilibrium inside the inner layer cable cross-section could be reached in about 70 μs by thermal conduction inside the copper without help of He. Therefore reaching the equilibrium will take longer for any practical cable and this effect can also be neglected during an "instantaneous" loss.

3.1.3 <u>Between cable and surrounding Helium</u>

At He II temperature the heat transfer between copper and surrounding helium has a time constant in the range of 25-60 μs. Then the heat transfer time from Cu to He II is comparable with the transfer between strands inside a perfect soldered cable.

---

* Du Pont trade-mark of a polyimide film

3.2 <u>During a "continuous" loss</u>, it is assumed that the beam losses produce a uniform temperature in any cross-section of the cable. But the temperature can change along the cable length, depending on the local energy deposit and heat exchange conditions.

Equilibrium temperatures are reached along the chain of thermal resistances. These temperatures must be low enough to maintain the SC cable anywhere below the critical current temperature.

In case of cooling by superfluid Helium, the maximum temperature should not exceed the critical Helium temperature (2.17 K) in order to keep the benefit of its high thermal conductivity and low viscosity.

Experiments are going on at Saclay[5] to measure the difference of temperature between the insulated cables and a static bath of superfluid Helium, when the cables are heated. A sample of SC coil is simulated by five stainless steel pseudo-conductors machined to have the same external surface as a SC cable. Temperature sensors are placed in contact with the metal. The conductors are insulated and resin polymerization is made under pressure as the SC cables are. Each of the three central conductor can be electrically powered. The cable and Helium bath temperatures are recorded versus the dissipated power in the conductors.

To keep some margin, a temperature difference of 0.15 K can be tolerated between the cables and the Helium bath. Such a temperature difference has been measured for a dissipated power of ~ 10 mW/cm^3. This power can change by a factor 2, according to the pressure during polymerization and to geometrical parameters of the tapes. Systematic studies are being made to understand the influence of these parameters and to fit the results with a thermal model. The Kapton thermal conductivity versus its thickness and the Kapitza resistance between Kapton and superfluid Helium have been measured.

4. QUENCH LIMIT

Dividing the maximum power which can be dissipated in the coil to maintain it at a reasonable temperature (Ch. 3), by the maximum energy deposit per incident proton in the coil (Ch. 2), gives the number, N_p, of lost protons per dipole corresponding to the quench threshold for a continuous loss.

For the LHC, the maximum energy deposit per incident "gaussian" proton is around 3 Gev/cm^3 (= $4.8*10^{-10}$ J/cm^3); a dissipated power of ~10 mW/cm^3 (1.2 mW/g) creates a maximum tolerable temperature difference of 0.15 K. The quench threshold for a point loss is then :

$N_p < 2*10^7$ p/s

For the SSC[6] it is assumed that the maximum energy deposit per incident proton in the coils is 50 Gev/cm^3 and the power required to quench a magnet is about 10 mW/g. The quench threshold for a point loss is:

Np < $1*10^7$ p/s

The nominal quenching limits of the Tevatron[7] are 8 mW/g for slow losses and 1 mJ/g for a fast loss.

Experimental verification with beam in SC machine is not easy to be done. The number of lost protons is difficult to be measured as well as the power deposit.

Some experimental results have been obtained at FERMILab, to determine the quench threshold due to beam loss in dipoles and quadrupoles during fast resonant extraction (~1 ms)[8]. Elastic particles escaping from the septum are lost all around the accelerator. Particle tracking, taking into account the geometry of the vacuum chamber, gives a loss distribution, which fits quite well with the loss distribution measured by the beam loss monitors. Near a selected dipole, the beam is moved by a closed 4-bump so as to enhance the losses of secondary particles and the beam intensity is raised by step until the element quenches.

The tracking gives a longitudinal "hit" distribution of the secondary particles striking the dipole. For each hit a hadron/electromagnetic cascade plus elastic scattering code computes the energy deposit in the coil. The summation of all the hits is made. The quench is obtained for a maximum energy deposit per incident proton of 1.2 Gev/cm^3 leading to a maximum energy deposit of 8.8 mJ/g with $3.6*10^8$ particles striking the dipole. This number, given with significant statistical error, corresponds to a transient state,(1 ms), where the local Helium enthalpy improves the situation regarding to a continuous loss.

5. COLLIMATOR EFFICIENCY: EDGE EFFECT, IMPACT PARAMETER

Energy deposition in magnets from beam losses in interaction regions are not cause for alarm in today's high energy p-pbar colliders, since the luminosity, even in the case of the upgraded Tevatron[9], does not exceed 10^{31} cm^{-2} s^{-1}.

On the contrary, high-energy p-p colliders with luminosity exceeding 10^{33} $cm^{-2}s^{-1}$ are faced to the problem of having a very efficient collimator system.

For the SSC[6], assuming $4*10^{14}$ protons and a 300 hr total lifetime, the loss around the ring is $3.7*10^8$ p/s. For the LHC the number of particles lost outside the insertions exceeds 10^9 p/s.

As the losses could be concentrated on the single dipole where the smallest aperture restriction appears, and as they should not exceed 10^7 p/s per dipole, the efficiency of a collimator system should be better than 99 %.

5.1 Edge effect

High energy particles striking beampipes, scrapers, collimators, close to the surface and nearly parallel to it, have a significant chance to be reflected back out of the target.

They only participate to elastic and quasi-elastic processes. This collimator "edge" effect has been studied by A.van Ginneken[10,11]. It drastically reduces the efficiency of a single collimator, while the escaping particles close in energy and angles to the beam travel with the beam for considerable distances.

Fig.4 to 6 show the importance of the "impact" parameter of the incident proton, i.e. the angle of incidence, x'_0, and the initial position relative to the edge, X_0. As Coulomb and nuclear scattering are the dominant processes, angles and distances scale with momenta as p^{-1} and the figures can be extrapolated in the range 1-20 Tev. The efficiency of a collimator system can only be known if the impact parameters are defined.

5.2 Emittance growth

Once a collimator is set to be a restricted aperture by intercepting the beam distribution, say, at 6σ, the new primary protons striking the collimators come from the emittance growth; secondary particles escaping from the collimator can also strike again the collimator after several turns. The impact parameters are then determined only if the effects producing the emittance growth are quantitatively known, which is not often the case!

An experiment was launched recently on the CERN Sppbars[12]. A collimator jaw can be set in the vacuum chamber, all the others collimators being retracted, and a scintillator, installed an integer of half betatronic wavelengths downstream, can be moved into the vacuum chamber. When the scintillator is set in the "shadow" of the collimator, the particles escaping from the collimator can hit the scintillator depending on their emission angle and the relative position of the scintillator. The signal output of the scintillator is detected by a photomultiplier and recorded at each revolution of a pre-selected bunch.

Position pickups give the closed orbit at the collimator and scintillator locations; the beam emittance is computed from transverse beam size measurements. The collimator is set at a few σ of the Closed Orbit axis and the scintillator some more σ outside. The collimator is then removed as fast as possible and the evolution of the scintillator rate is recorded. (Fig.7). The experiment is repeated for different positions of the collimator and of the scintillator relative to the beam axis.

Due to non-linearities in the machine, it is expected that a proton escaping from a bunch gets a velocity, $V(Y)$, increasing with its distance to the center of the bunch, Y_0. A model like $V(Y) = a (Y - Y_0)^n$ has been checked by fitting the experimental data. The result (Fig. 8) shows a sharp increase of the average transverse drift speed when the particles are around 8-9 σ.

A rough approximation of the impact parameter, X_0, can be deduced. Assuming the collimator is set at 6σ, the drift speed given by Fig.8 is 5σ / s ~ 5 mm/s in the SPS at 315 Gev/c. The

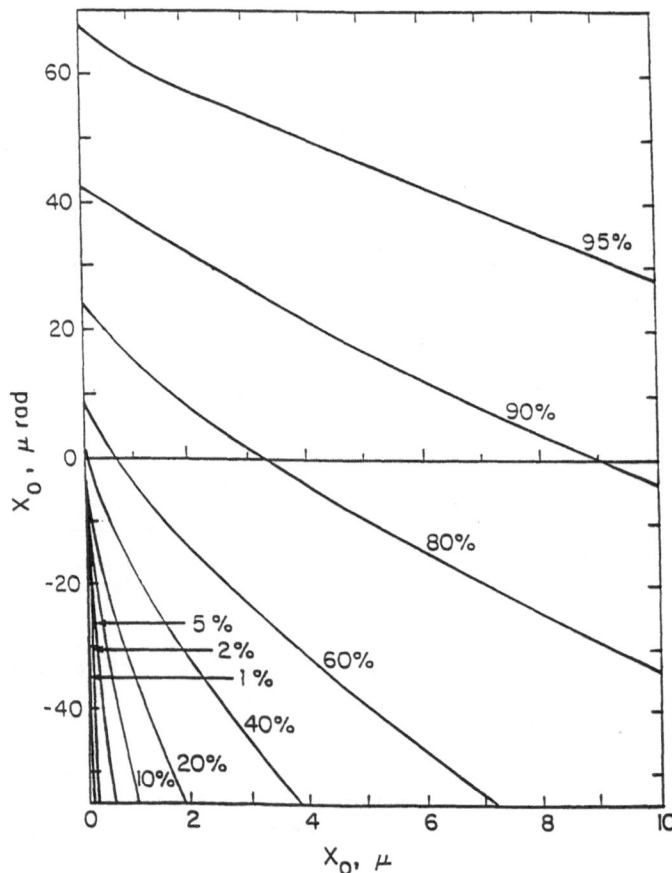

Fig.4   Iso-absorption contours for infinitely long, semi-
        infinitely wide iron target for a 1 TeV/c pencil beam
        of protons as a function of displacement x_0 and angle
        x'_0 of the beam.[11]

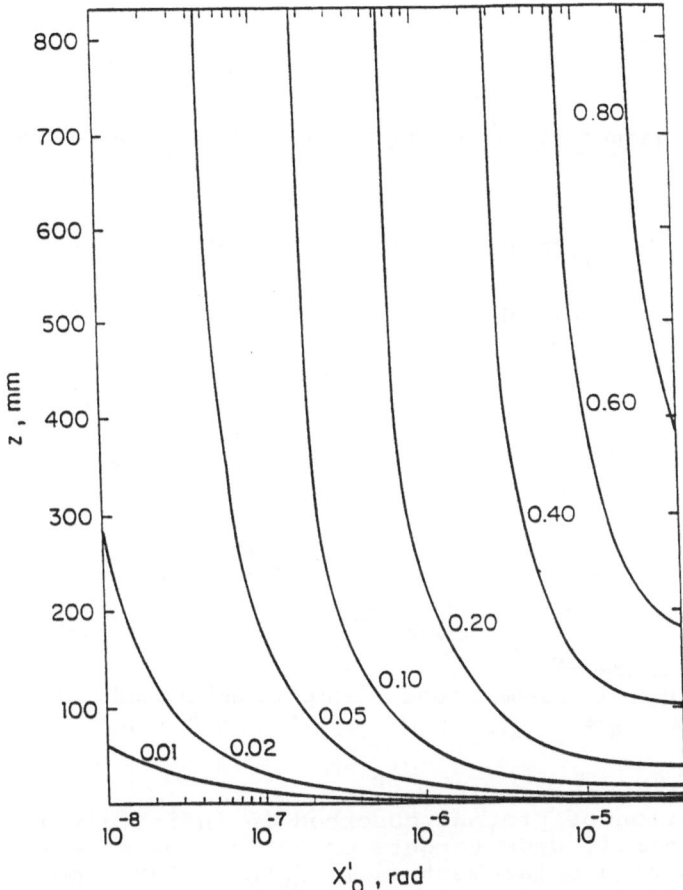

Fig.5   Iso-absorption contours for semi-infinitely wide iron
target for a 1 TeV/c pencil beam of protons incident
on edge of target as a function of angle (x'_0) and
target length (z). [11]

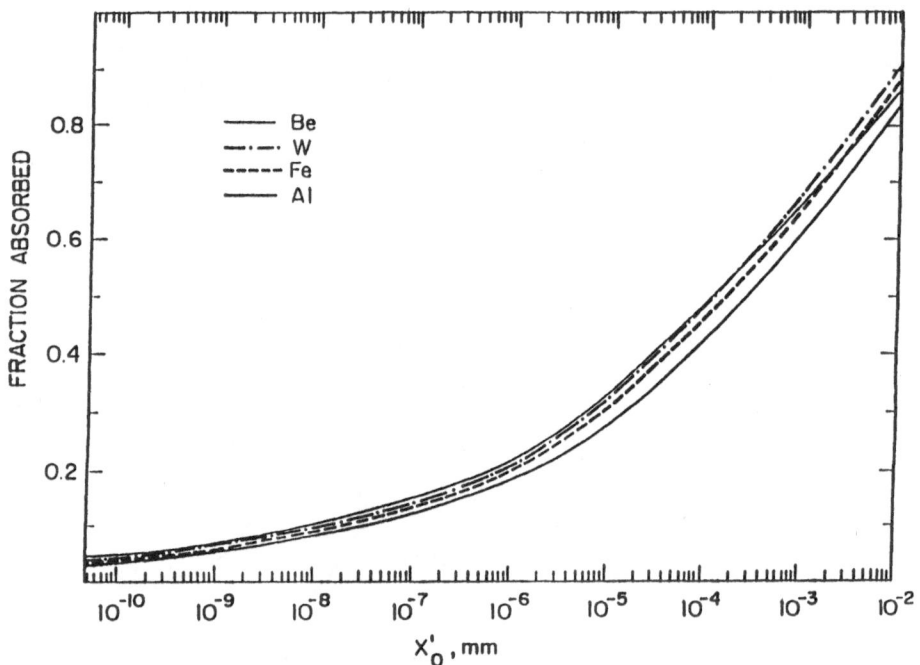

Fig.6 Fraction of protons absorbed in infinitely long, semi-infinitely wide targets of various materials as a function of displacement (x_0) of the 1 TeV/c pencil beam parallel to edge of target. [11]

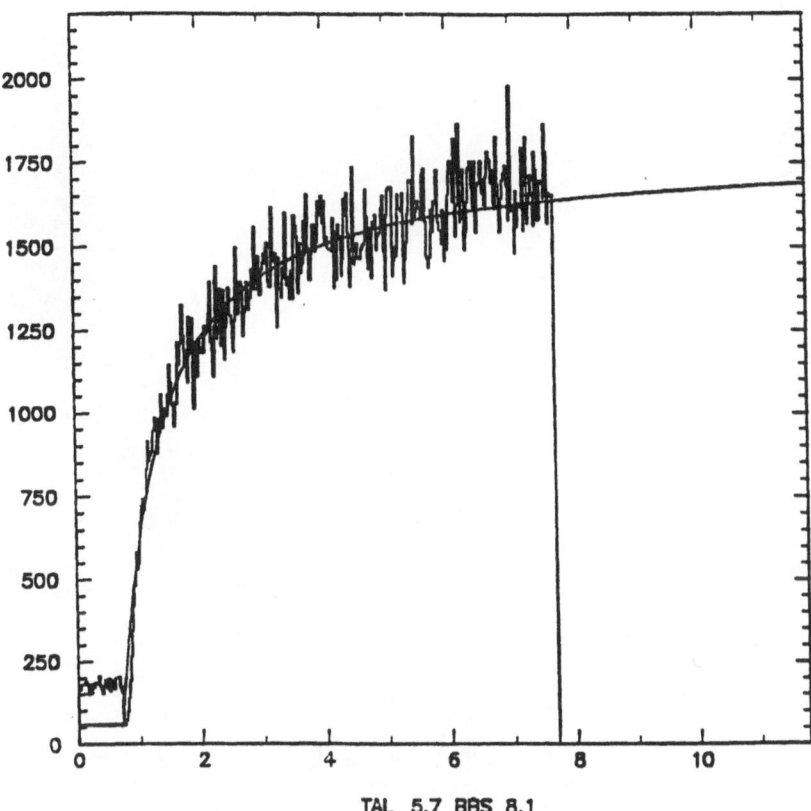

TAL 5.7 BBS 8.1

Fig.7 Scintillator rate evolution when the collimator is
retracted from 5.7 σ/beam axis up to full aperture.
Scintillator position: 8.1 σ /beam axis.

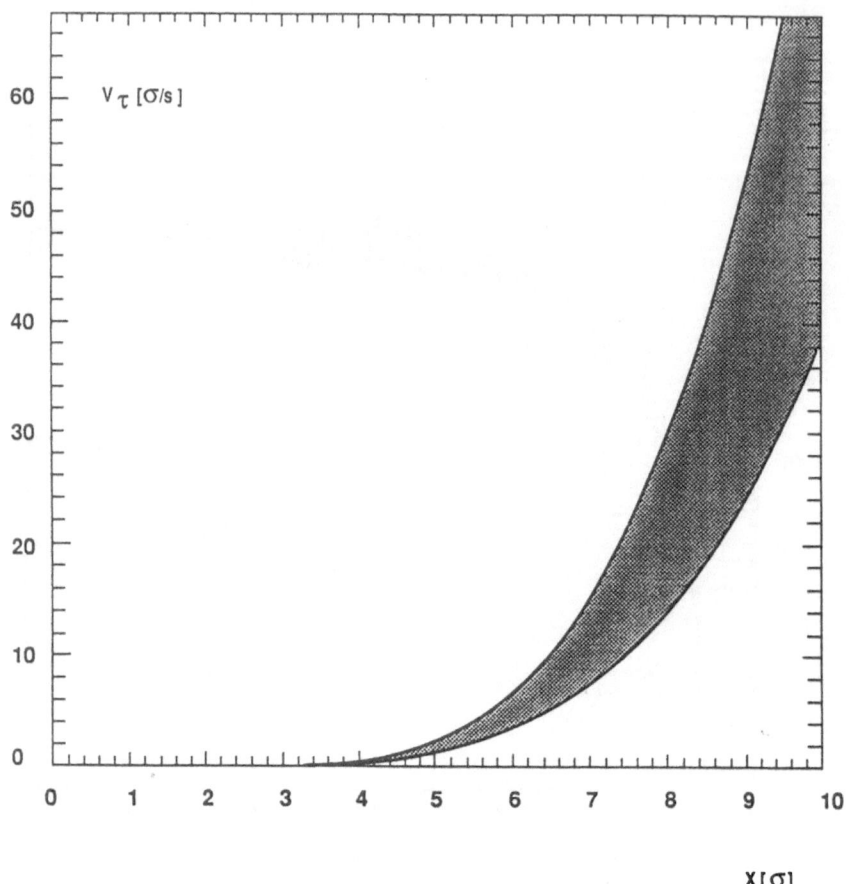

Fig.8   The average transverse drift speed as a function of
        the distance to the beam axis.

average drift speed per SPS revolution is ~ 0.1 μm. Assuming such a regular spiral increase, it will take in average 50 turns for a proton located anywhere at 6σ in the phase space to reach the collimator. 50 % of the protons hitting the collimator will have an impact parameter $X_0 < 5$ μm. Other effects like the smear, not yet known quantitatively, can modify this impact parameter by adding a random component from one revolution to the other.

CONCLUSIONS

The energy deposit in SC coils obtained from theoretical simulations and validated by some experiments, limits to a few 10^7 p/s the continuous flux of lost protons per magnet before quenching.

If the loss occur in a relatively short time, a number of 10^8 to 10^9 p seems to be tolerated before quenching, but transient phenomena have to be studied more carefully.

Anyway these numbers represent a tiny fraction of the circulating beams ($> 10^{14}$ p). To avoid catastrophic failures due to a full beam loss, new control methods should be envisaged. Real-time controls with automatic feedback between beam loss monitors and beam parameters could be tried to avoid frequent beam dumps and the lengthy procedure of injection and acceleration, which reduces the available time for Physics.

For high energy colliders with a luminosity exceeding 10^{33} cm^{-2} s^{-1} sophisticated systems should be designed to absorb the continuous halo. A single collimator with a perfect "flat" edge and perfectly parallel to the incident particles is not efficient enough due to the edge effect. Furthermore the impact parameters of incident particles are of the same order as the irregularities of the "flat" edge, and, if not perfectly parallel to the beam, a collimator is traversed by the halo only at its upstream or downstream corner, reducing the interaction region. The first obstacle to be hit by the halo cannot act as an absorber, but only as a scraper increasing the angle of the escaping particle relative to the circulating beam. Taking into account expected beam parameters and particle tracking, a possible solution could be a combination of collimators ending with an absorber in a long straight section.

REFERENCES

1. L. Burnod, D. Leroy, Influence of beam losses on the LHC magnet, LHC Note 65, CERN, December 1987.
2. L. Burnod, J.B. Jeanneret, Beam losses in the SPS and the LHC due to beam-gas and beam-beam collisions, LHC Note 91, CERN, April 1989.
3. A. van Ginneken, CASIM, FERMILab TN-309 (1978).
4. P. Aarnio et al., FLUKA, CERN-TIS-RP/168, January 1986.
5. C. Meuris, Essais thermiques des échantillons LHC, CEA/IRF/DPhPE/STCM, TC 2, April 1989.
6. M.G.D. Gilchriese, Report of the task force on Radiation levels, SSC-SR-1035, p 395.
7. F.T. Cole et al., Report of the design of the FNAL SC Accelerator, FERMILab. (1979).

8.   A. van Ginneken, D. Edwards, M. Harrison, Quenching induced by beam loss at the Tevatron, FERMILab-Pub-87/113, July 1987.

9.   A. van Ginneken, Energy deposition in Tevatron magnets from beam losses in Interaction Regions, FERMILab TM-1548, October 1988.

10. S. Qian, A. van Ginneken, Energy deposition in large targets by 1-20 TeV Proton Beams, FN-514, F.N.A.L., May 1989.

11. A. van Ginneken, Elastic scattering in thick targets and edge scattering, FERMILab-Pub-87/141, F.N.A.L, Sept. 1987.

12. L. Burnod, J.B. Jeanneret, Drift speed measurements of the halo in the SPS collider, LHC Note 117, CERN, January 1990.

Annex 1 - BEAM-GAS INTERACTIONS[2]

The beam passing through the residual gas is scattered by Multiple Coulomb Scattering (MCS) or creates Nuclear Interactions (NI). Both depend on the gas density.

In a superconducting machine like the LHC, almost all the vacuum chamber is at liquid Helium temperature, 4.5 K. The residual pressure, measured with a room temperature connection, is expected to be at most $3 \cdot 10^{-12}$ Torr mainly due to Hydrogen if there is no leak. This pressure corresponds to an atomic density in the vacuum chamber $(N_a)_{4.5°K} = 1.73 \cdot 10^6$ atoms/cm^3

The target length, in g/cm2, equivalent to the residual gas is given by $x = \rho \cdot L$, where ρ is the gas density ($\rho = 3.0 \cdot 10^{-18}$ g/cm3) and L the target length in cm (one machine-turn integrated density: $L = 2.67 \cdot 10^6$ cm). The number N_c of atoms/ cm2 corresponding to the target length is $N_c = N_a \cdot L = 4.8 \cdot 10^{12}$.

1. MULTIPLE COULOMB SCATTERING (MCS)

The r.m.s MCS angle per turn projected on one plane is given by :

$$< \Theta >_{MCS} = (15 \cdot 10^{-3} \cdot (x/X)^{1/2})/p$$

where

p = particle momentum in GeV/c = 8000 GeV/c
x = target length in g/cm^2 = $7.9 \cdot 10^{-12}$ g/cm^2 over

one LHC turn

X = radiation length = 63 g/cm^2 for Hydrogen
$< \Theta >_{MCS}$ /turn = $0.66 \cdot 10^{-12}$ rad

The Multiple Coulomb Scattering progressively increases the initial emittance. After one day of storage (86400 s), the average MCS angle is

$$< \Theta >_{day} = < \Theta >_{turn} \cdot (Frev \cdot 86400)^{0.5} = 2.06 \cdot 10^{-8} \text{ rad}$$

The MCS is a continuous process, so the average angle can be compared with the standard deviation angle σ' at beta average, due to the beam emittance ε^*:

$\sigma' = (\varepsilon^* / (\gamma \cdot < \beta >)^{0.5}$ at the beginning of a coast.
After one day, σ' becomes
$\sigma'_{end} = (\sigma'^2 + <\Theta>_{day}^2)^{0.5}$

For the LHC, $< \beta > = 100m$ and with the minimum beam emittance $\varepsilon^* = 5 \ \pi.mm.mrad$

$\sigma' = 2.42 \cdot 10^{-6}$ rad, $< \Theta >_{day} = 2.06 \cdot 10^{-8}$ rad and $\sigma'_{end} = 2.42 \cdot 10^{-6}$ rad

or a relative increase of $3.6 \cdot 10^{-5}$ after one day!

So, at least for LHC, the MCS will not induce significant beam losses by itself.

2. NUCLEAR INTERACTIONS

For Hydrogen, in the LHC at 8000 Gev/c

-the approximate center-of-mass energy $(s)^{1/2} = (2 \cdot p \cdot A \cdot E_0)^{1/2} = 122.5$ Gev/c

where p is the beam momentum (Gev/c), A the atomic number of the target, E_0 the proton rest energy (= 0.938 Gev)
  -the proton-nucleon total cross-section, $\sigma_{TOT} = 49$ mbarn
  -the proton-nucleon elastic cross-section, $\sigma_{ELA} = 9$ mbarn
  -the total (λ_{TOT}) and elastic (λ_{ELA}) collision lengths defined by $\lambda = (A \cdot m_p / \sigma)$ are $\lambda_{TOT} = 34$ g/cm2 and $\lambda_{ELA} = 186$ g/cm2
  -the elastic differential cross-section defined in the usual way as

$d\sigma_{ELA}/dt = a \cdot e^{(-b \cdot t)}$, where $t = (p \cdot \Theta)^2$, $1/b = (p \cdot \Theta_0)^2$ and $\sigma_{ELA} = a/b$

(t: 4-momentum transfer, Θ: polar angle)
a = 126 mb/$(Gev/c)^2$, b = 14 $(Gev/c)^{-2}$, $\Theta_0 = 0.031$ mrad

The total mean free path between 2 nuclear collisions (mfp_{TOT}) and between 2 nuclear elastic collisions (mfp_{ELA}), defined by $mfp = \lambda/\rho$ are

$mfp_{TOT} = 1.1 \cdot 10^{19}$ cm or $4.2 \cdot 10^{12}$ LHC turns or $3.8 \cdot 10^8 s = 4300$ days
$mfp_{ELA} = 6.2 \cdot 10^{19}$ cm or $2.3 \cdot 10^{13}$ LHC turns or $2.0 \cdot 10^9 s = 23500$ days

They are very large for the low gas densities in the LHC case.

With the same gas and densities and the cross-sections given above, the luminosity for the beam-gas nuclear interactions ($L = N_c \cdot N_p \cdot f_{rev}$) and the corresponding number of events are

N_p/bunch	$1.0 \cdot 10^{11}$
Nr of bunches / beam	4875
N_p / beam	$4.87 \cdot 10^{14}$
L / beam (cm-2*s-1)	$26 \cdot 10^{30}$
Nuclear events / beam*s	$13 \cdot 10^5$
Elastic events /beam*s	$2.3 \cdot 10^5$
Tn (s)	$3.8 \cdot 10^8$
Tn (Hrs)	104000

Protons having got an inelastic collision produce secondary particles with much lower energy. Those particles are stopped in the nearby bending magnets, just downstream of the location of the interaction. The losses being distributed everywhere, a localized collimator has no effect. But the average energy deposited by meter of bending magnet is small:

~ 49 protons /m · s = 63 µw/m for the LHC

The nuclear interaction of the beam with the residual gas leads to an exponential decay of the beam intensity such as $I = I_0 \cdot e^{-t/Tn}$ or $dI = -I \cdot dt/Tn$, Tn being the nuclear proton lifetime due to the gas. Assuming all the protons having got a nuclear interaction are lost, the lifetime due to nuclear interactions, Tn = I*dt/dI , is very large for the LHC, where the nuclear interactions on the residual gas can then be neglected as far as the lifetime is concerned.

STATUS REPORT ON SSC DIPOLE R&D[*]

C.Goodzeit, P.Wanderer, E.Willen, J.Cottingham,
G.Ganetis, M.Garber, A.Ghosh, A.Greene, R.Gupta,
J.Herrera, S.Kahn, E.Kelly, G.Morgan, J.Muratore,
A.Prodell, M.Rehak, E.P.Rohrer, W.Sampson, R.Shutt,
P.Thompson

Brookhaven National Laboratory, Upton, N.Y.,U.S.A.

M.Chapman, J.Cortella, A.Desportes, A.Devred[1],
J.Kaugerts, T.Kirk[2], K.Mirk, R.Schermer[3],
J.C.Tompkins,J.Turner

Superconducting Super Collider Laboratory, 2550
Beckleymeade Ave., Dallas, TX 75237,U.S.A.

M.Bleadon, B.C.Brown, R.Hanft, M.Kuchnir, W Koska,
M.Lamm, P.Mantsch, P.O.Mazur, D Orris, J.Strait,
G.Tool[4]

Fermi National Accelerator Laboratory, Batavia, Il
60510, U.S.A.

R.Althaus, S.Caspi, W.Gilbert, C.Peters, J.Rechen,
J.Royet, R.Scanlan, C.Taylor, A.Wandesforde,
J.Zbasnik

Lawrence Berkeley Laboratory, Berkeley, CA 94720,
U.S.A.

Now at:

[1]KEK, National Laboratory for High Energy Physics,
1-1 Oho, Tsukuba-shi, Ibaraki-ken 305 Japan

[2]Argonne National Laboratory
Argonne, IL 60439, U.S.A.

[3]Lawrence Berkeley Laboratory
One Cyclotron Road, Berkeley, CA 94720, U.S.A.

[4]Superconducting Super Collider Laboratory
2550 Beckleymeade Ave, Dallas, TX 75237, U.S.A.

---

[*]Work supported by U.S. Department of Energy

INTRODUCTION

This paper describes the principle design features of the main collider dipole magnet that is being developed for the Superconducting Super Collider (SSC). The magnet is a collaborative effort between the SSC Laboratory, Brookhaven National Laboratory, Lawrence Berkeley Laboratory and Fermi National Accelerator Laboratory.[1] The design is based on a high central field (6.6 T) over a small aperture (40 mm coil ID) and long (16.6 m) effective length. This paper describes the features of the current design and also reports quench field and field uniformity data from recent magnets. Test results from earlier short and long models have been reported elsewhere.[2] Figure 1 shows a cross section of the collared coil and Fig.2, a cross section of the cold mass.

COIL DESIGN AND CONSTRUCTION

The coil design[3], known as C358D, includes three copper wedges in the inner coil and one copper wedge in the outer coil. Both coils are wound from a partially keystoned cable of the Rutherford type, with high homogeneity NbTi. The wedges compensate for the partial keystone angle of the cable and furnish additional, required degrees of freedom in the field shaping optimization procedure. To ensure cost-effective super-conductor utilization, the inner coil (16 turns) is wound from a 23-strand (wire size 0.808 mm) cable of Cu:SC ratio 1.3:1 and the outer layer (20 turns) from a 30-strand (wire size 0.648 mm) cable of ratio 1.8:1. The filament size is 6μm. The cable thickness is controlled at the specified value with a variation of less than 6.3μm. Cable insulation consists of a double layer of 25μm Kapton followed by a layer of epoxy-impregnated (23% by wt.) fiberglass. The coil ends maintain the same radial dimensions as in the straight section and are wound in a constant-perimeter configuration. Spacers between groups of turns in the ends shape the field for low multipole content and reduced peak field. Between individual turns, they also protect against electrical shorts. Fiberglass cloth, knife-coated with mineral-filled epoxy, is used on the radial surfaces of the ends to add strength and to give a constant radial thickness throughout. Coils are cured under pressure (~7 kpsi) and temperature (~140 C) in a precision mold to give a rigid structure that can be easily handled and that has precise dimensions. The variation in azimuthal dimension of the coils is less than ±50μm peak-to-peak and less than 35μm RMS. These rigorous tolerances ensure a constant collaring stress as well as acceptable multipole variation.

The coil design is such that the inner layer will quench first with increasing current. The specified current density for the wire of the inner coil is 2750 A/mm^2 at 4.22 K and 5 T. Allowing for 10% current degradation in cable manufacture, the C358D design central field is expected to reach 6.86 T at the SSC operating temperture of 4.35 K before quench. Thus,there is a 4% field margin in the design of the magnet. In practice, short model magnets have generally reached fields a little higher than expected from short sample measurements on the cable, at currents indicating that the superconductor can, under some conditions, operate somewhat higher into the resistive region than 10^{-14} ohm m before quenching. Many of the model magnets built to date have utilized a higher Cu:SC ratio of 1.5:1 for the inner cable because of concerns about magnet stability with the specified 1.3:1 value.

COLLARS AND COIL PRESTRESS

The coils are compressed with 15 mm wide collars of fully austenitic, nonmagnetic stainless steel. The collars supply a compressive pressure of

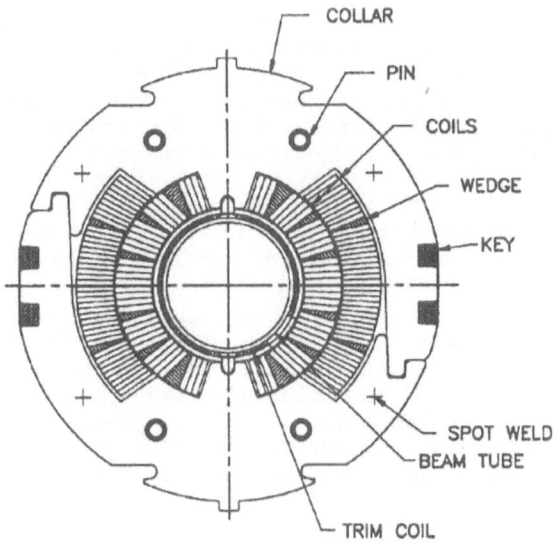

Fig.1 A cross section of the collared coil assembly

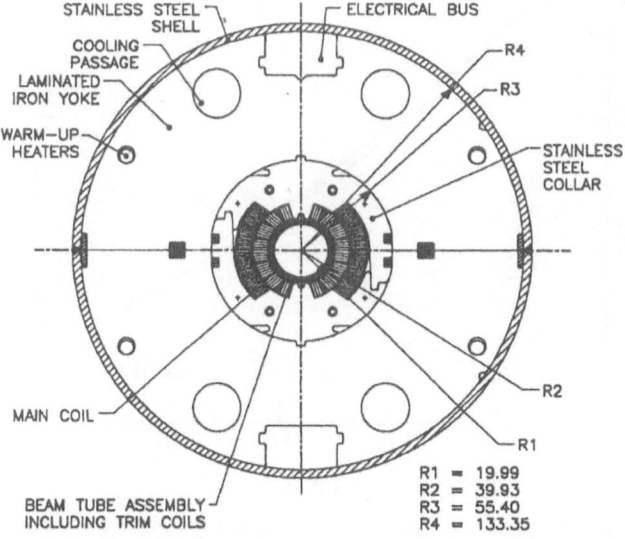

Fig.2 A cross section of the cold mass assembly

~8 kpsi, sufficient to restrain coil motion under the action of the Lorentz force at high field.The collars are spot-welded in pairs for greater stiffness and, being asymmetric in shape, are alternated between left and right shapes as they are assembled into packs (15 cm long) in order to avoid the introduction of twist into the collared structure. Brass shims are applied to the collar packs at the poles to adjust for varying coil sizes in the R&D program. The compression of the collars is done in a hydraulic press in which, after sufficient vertical force is applied, a horizontal force pushes the full-hard phosphor bronze tapered keys (3° taper) into position. To reduce the required horizontal force and to reduce scoring of the keys, a film of lead-based lubricant is applied to the keys before installation. Upon release from the press, there is typically a loss of 2 kpsi in the inner coil stress from the maximum experienced during the keying operation. The diameter of the collared assembly typically grows by 0.25 mm vertically from the unstressed state; the horizontal diameter remains nearly unchanged.

In order to determine the azimuthal stress on the coils, not only during the collaring operation but also during cool-down and during excitation of the magnet, a strain gauge system has been developed[4] that is capable of giving accurate readings of the stress under these varying conditions. It consists of precision EDM-cut stainless steel bars that respond to force by bending as simple beams and to which are attached strain gauges responding to the beam elongation. These transducers are mounted in a special collar pack with one at the pole of each coil section for a total of eight transducers in one pack.

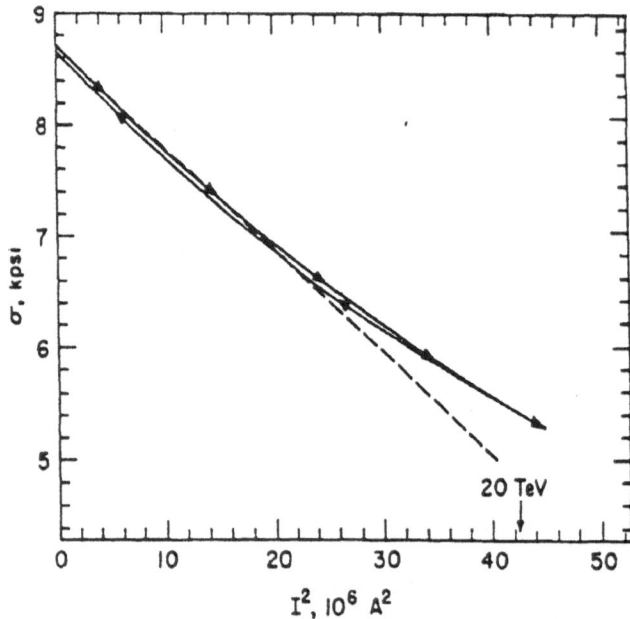

Fig.3 A measurement of coil stress vs. current in 17 m magnet DD0017. The data are the lines with arrows

An electronic system using an accurate current source and a high-precision integrating voltmeter is utilized for individual readout of each transducer, thereby avoiding the difficulties of bridge circuits frequently utilized for this purpose. Compensating transducers free of stress are included in each package to adjust for thermal and magnetic field effects. Figure 3 shows a measurement[5] of coil stress vs. current made with this strain gauge system. These data show the expected stress variation with the square of the current and the desired residual stress in the coils at high current.

The coils and collars are assembled around a copper-plated beam tube with space for multipole trim coils to be mounted on the surface[6]. Notches in the collars at top and bottom engage tabs mounted on the beam tube to provide precision alignment.

IRON YOKE

The yoke laminations (Fig.2), punched from low-carbon steel of thickness 1-1/2 mm (future models will use steel of thickness 6 mm) and assembled into blocks 15 cm long, are designed to support the collared coil around the circumference. This added restraint of the Lorentz force was not a feature of earlier designs of the SSC magnet. The collared coil is oriented inside the iron yoke with tabs at the top and bottom. The top/bottom yoke halves are aligned via keys on the horizontal midplane. The two large rectangular slots carry bus work for a magnet string and the four large round holes are passages to carry helium through the magnet system. Holes are provided in the yoke for heaters, required in the machine to warm-up a sector of magnets in the specified 24 hours. Upon assembly, the yoke midplane gap is no more than 0.05-0.075 mm; this gap is closed when the outer stainless steel shell that captures the yoke and provides the helium containment is welded in place. Flats are provided on the perimeter of some yoke laminations for alignment of the cold mass structure. Heavy end plates (3.8 cm thick) coupled to the shell prevent significant end motion under the axial Lorentz force. A strain gauge system monitors these forces.

The iron contributes ~1.7.T to the 6.6 T central field of the magnet. With the 15 mm separation of the coil from the yoke provided by the collars, saturation effects are quite small: the transfer function decreases by ~3% and the sextupole component changes by $< 0.4 \times 10^{-4}$ (at 1 cm) due to iron saturation.

ELECTRICAL INSULATION

The high dielectric strength, polyimide film Kapton is used to insulate the coils of the magnet against turn-to-turn and turn-to-ground voltage breakdown. Such electrical failures can occur because of puncture through the film or by flashover around the edge of the film. Puncture-type failure may occur if the film contains manufacturing defects (pinholes) or damage. Failure due to pinholes is avoided by using multiple layers of material, thereby giving a low probability that pinholes will align to cause failure. The Kapton used to construct magnets is inspected to ensure less than 10 pinholes/m^2. Damage to Kapton is avoided by good construction practice although it can occur if local pressures are excessive. An R&D program is underway to identify filled Kapton material that is less susceptible to the plastic creep that can lead to pressure failures. Flashover is avoided by the design practice of maintaining creep paths of >5 mm from any conductor around film edges to ground. This allows coil testing ("hipotting") in air at the desired value of 5 kV while maintaining the specification of 1 kV/mm common in electrical design work.

Voltages in the magnet coils during quench can reach 1500 V with respect to ground and turn-to-turn voltages can reach 70 V, based on calculation. As standard practice for SSC magnets, ground insulation integrity is checked by hipot testing at twice the expected voltage plus 2000 V, or 5000 V total. This testing is done under normal room conditions. Turn-to-turn insulation is checked by applying a 2000 V voltage pulse to each coil after collaring and looking for deviations from the expected ringing pattern of the pulse. This results in a per turn test of greater than 70 V but not quite the factor of two test that would be desirable; higher voltage is not used because of the desire to avoid excessive terminal voltage. These tests are carried out repeatedly throughout the construction process to ensure the electrical integrity of the magnet.

MAGNET COOLING

A modification to the original cooling scheme for the magnets has been adopted which forces the specified 100 g/sec of supercritical helium (4 atm pressure) to circulate from the top passages in the iron yoke around the coil and beam tube to the bottom passages. This is accomplished by blocking the top and bottom passages at alternate ends of the magnet, thereby forcing the helium to pass through slotted, stainless steel laminations placed at regular intervals in the yoke, along the collar loading flats, then between collar packs to the annular space between the bore tube and the coil ID. The helium is forced to flow inward between collar packs by periodic blocks placed in the collar loading slots. The original cooling scheme depended on conduction of heat from the bore tube outward through the magnet components to the passages; only 1 gm/sec circulated in the annular space between beam tube and coil, serving primarily to transfer heat between the two but not in itself able to extract much heat. Heat is deposited inside the beam tube due to synchrotron radiation; at 20 Tev, this amounts to ~2 watts at the design luminosity but would reach higher levels for beam currents exceeding design. In addition, the warm finger required during testing to map the field deposits considerable heat in the bore tube. With the revised cooling scheme, 10 watts of heat deposited along the length of the magnet will result in a temperature rise in the magnet coil of only ~0.07 K; with the original cooling scheme, this temperature rise would be ~0.6 K. The pressure drop across the magnet remains suitably low at ~0.001 atm.

MAGNET QUENCH PERFORMANCE

Thus far, test results are available for six 17m dipoles built with improved end support and so that the collars are supported by the yoke (Fig.4). It can be seen that the magnets are at the limit of the conductor after zero to three training quenches. (The magnet-to-magnet variation in maximum quench current is due to variations in the conductor). This quench performance is substantially better than that of previous magnets, made with unsupported collars and much weaker end support. It was demonstrated in magnet DD17, which reached the conductor limit of 7.5 T at 3.5 K without further training, that the magnet construction has a significant safety margin with respect to operating at 6.6 T.

Magnets are also tested for their response to a thermal cycle to room temperature and back, as would be needed for accelerator repair. Some quench once or twice at currents slightly below the conductor limit before again reaching a plateau. The locations of quench origins are well known from voltage tap data, and work is underway to further improve performance.

As noted before, magnets built with conductor which has the specified current-carrying capacity and 1.3:1 copper-to-superconductor ratio will quench at currents 4% above the nominal operating current, 6.5 kA. Efforts are underway to increase this margin.

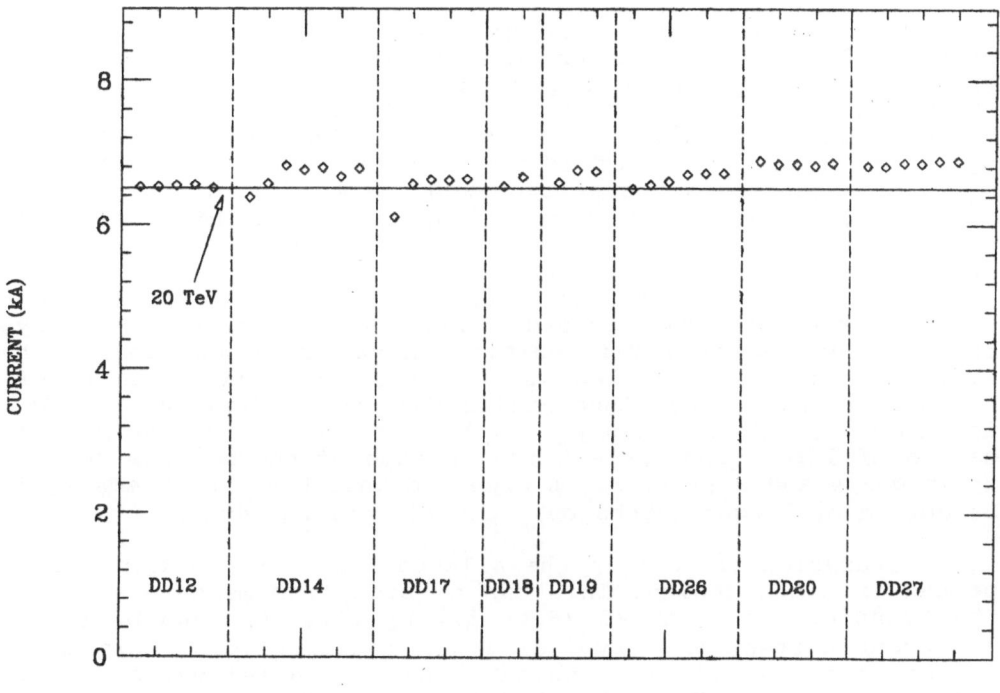

Fig.4 Initial quench history of recent 17 m SSC R&D dipoles

Due to the concentration of effort on improving quench performance, field quality measurements of 17 m magnets at operating temperature are just getting underway. However, considerable information exists from measurements of 1.8 m models and from room-temperature measurements of 17m magnets.

The 1.8 m models are not "field quality" magnets (made as identically as possible) since they are still being used as test beds for refinements in production. Thus, the focus is on understanding the relation between the measured field uniformity and magnet construction rather than on detailed comparison with SSC tolerances.

The standard expression for the multipole representation of the fields is:

$$B_y + iB_x = B_0 \sum_{n=0}^{\infty} (b_n + ia_n)(x + iy)^n$$

The multipole coefficients are evaluated at a radius of 1 cm, in dimensionless units of 10^{-4} of the dipole field.

The measured allowed and unallowed geometric coefficients of six 1.8m models of current design are given in Tables I and II, respectively. For the allowed terms, it is anticipated that it will be possible to reduce the mean values (e.g., b_4) by making small variations in the cross section of the coil. The most significant factor affecting magnet-to-magnet variation in b_2 is differences in the thickness of shims placed at the pole to control prestress. In production, this shim thickness will be held constant.

The unallowed terms are generally small but occasional excursions exist. The low-order unallowed terms are strongly affected by practical difficulties in magnet production, such as obtaining 17 m stainless steel yoke containment shells of sufficient uniformity. Asymmetries and the difficulty of welding these shells is thought to be the one of the principal causes of large quadrupole terms in some of the magnets (e.g., a_1 in DSS14 and DSS17). It is expected that these problems will be solved by larger purchases, special tooling, etc.

Saturation effects in the allowed terms are small, as expected from the design. This can be seen from measurements of the transfer function B/I (Fig. 5), b_2 (Fig. 6), and b_4 (Fig. 7) versus current in a typical 1.8 m model, DSV016. Warm measurements in 17m dipoles indicate that the axial variation of the dipole angle (Fig. 8), b_2 (Fig. 9), and b_4 (Fig. 10) is small.

Table I

COEFF.	DSS13	DSS14	DSS16	DSS17	DSK13	DSV16
b_2	-2.88	-1.80	-3.21	0.44	- .37	1.33
b_4	- .57	- .81	- .53	- .72	- .95	- .62
b_6	.02	.04	.03	.06	.07	.13
b_8	.05	.03	.04	.04	.03	.03
b_{10}	.07	.07	.06	.07	.07	.07
b_{12}	- .01	- .01	- .01	- .01	- .01	- .01
b_{14}	.00	.00	.00	.00	.00	.00

Table II

COEFF.	DSS13	DSS14	DSS16	DSS17	DSK13	DSV16
b_1	- .44	.61	- .73	-1.06	- .11	- .36
b_3	- .18	.02	- .22	- .13	- .01	- .11
b_5	.00	- .02	- .02	- .01	.01	.00
b_7	.00	.00	.00	.01	.00	- .01
a_1	- .69	-2.60	.09	-2.62	- .28	- .30
a_2	- .65	- .02	- .57	- .39	.04	- .63
a_3	.03	.06	.01	.44	.06	- .46
a_4	- .12	- .01	- .11	- .09	.03	- .12
a_5	- .02	- .03	.00	- .03	.00	+ .03
a_6	- .02	.02	- .02	- .03	.02	- .01
a_7	.00	.00	.00	.01	.00	.02
a_8	.00	.00	.00	.00	.01	0

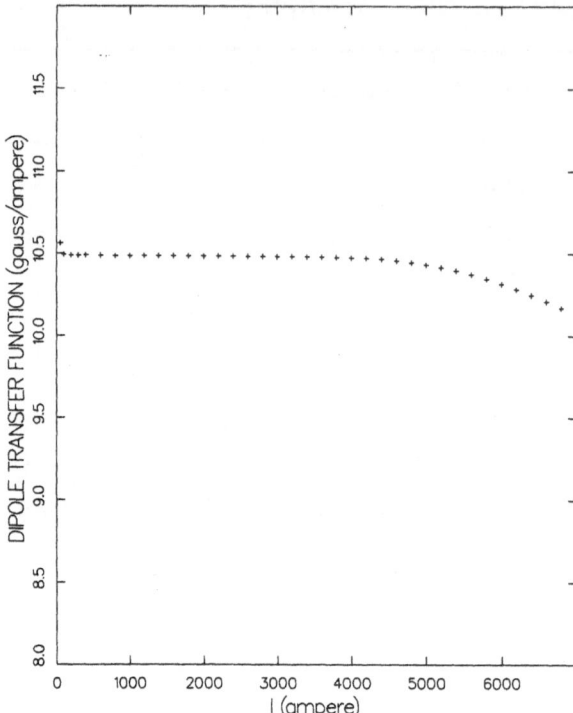

Fig.5 Transfer function (B/I) vs. I for a 1.8 m model dipole (DSV016)

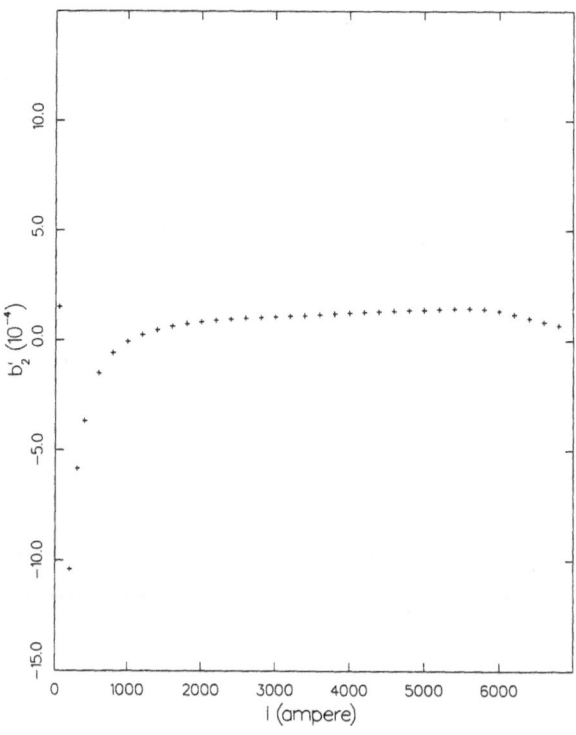

Fig.6 Sextupole coefficient b_2 vs. I for DSV016

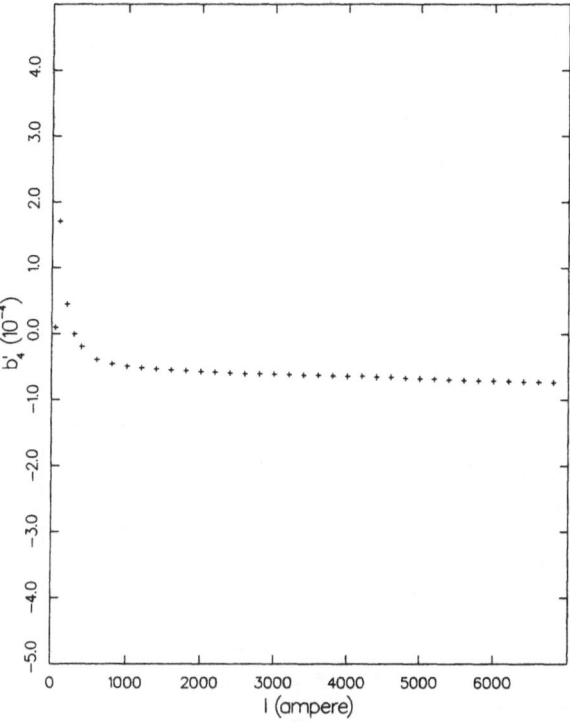

Fig.7 Decapole coefficient b₄ vs. I for DSV016

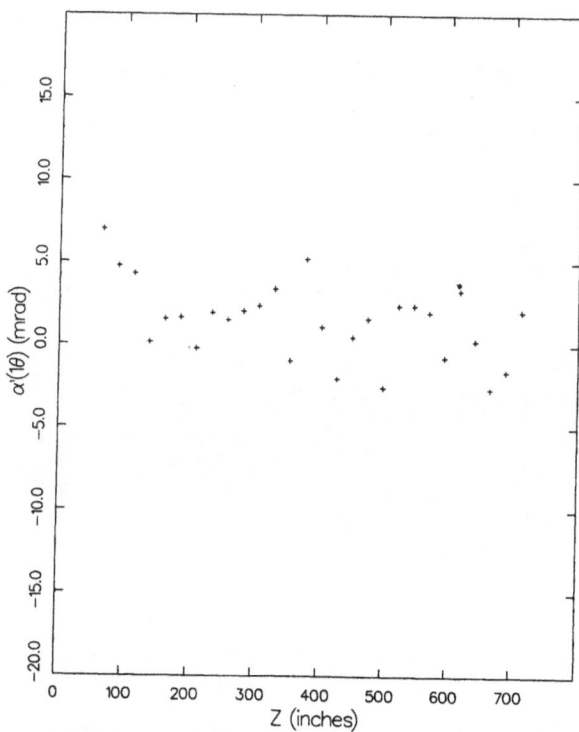

Fig.8 Axial variation of dipole angle in 17 m dipole
DD0017 at room temperature

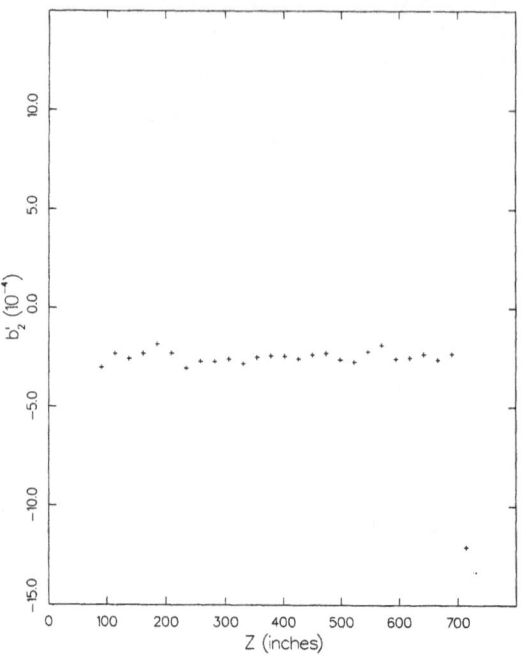

Fig.9 Axial variation of sextupole coefficient in DD0017 at room temperature

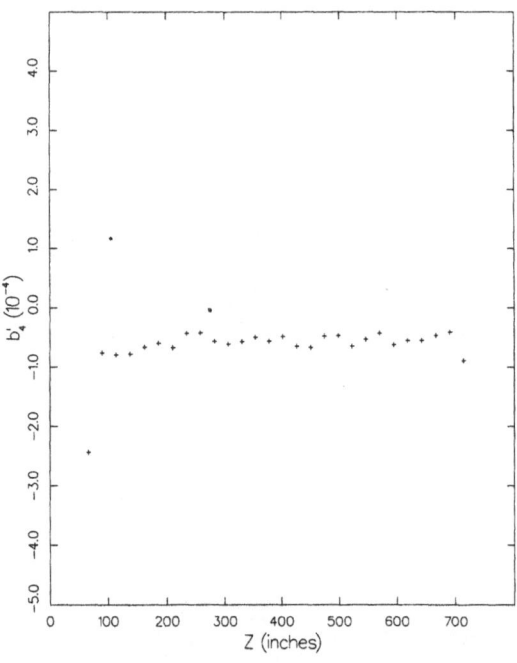

Fig.10 Axial variation of decapole coefficient in DD0017 at room temperature

REFERENCES

1. "Conceptual Design of the Superconducting Super Collider".
   SSC-SR-2020. SSC Central Design Group, Lawrence Berkeley
   Laboratory, One Cyclotron Rd., Berkeley, CA 94720 (1986).
2. E. Willen et al.,"Design Features of the SSC Dipole Magnet,"
   26th International Conference on High Energy Accelerators,
   Tsukuba, Japan (1989).
   P. Wanderer et al., "Test of Two 1.8 m SSC Model Magnets
   with Iterated Design," 1989 Particle Accelerator Conference,
   Chicago, IL (March, 1989).
   J. Strait et al., "Full Length SSC R&D Dipole Magnet Test
   Results," ibid.
3. G. Morgan. "C358A – A New SSC Dipole Coil and End Design."
   MDN-210-1 (1987),
   G. Morgan. "C358D – A Revision of the SSC Coil Design
   C358A." MDN-255-1 (1988).
4. C. Goodzeit, M. Anerella and G. Ganetis, "Measurement of
   Internal Forces in Superconducting Accelerator Magnets with
   Strain Gauge Transducers, "Proc.1988 Applied Superconduc-
   tivity Conf., IEEE Transactions on Magnetics, Vol.25, No.2,
   p.1463 (1989).
5. These data were taken at the FNAL magnet test facility.
6. J. Skaritka et al., "Development of the SSC Trim Coil Beam
   Tube Assembly", Proc. 1987 IEEE Particle Accelerator Conference, Washington, DC, p.1437 (IEEE, New York).

CRYOSTAT DESIGN FOR THE SUPERCONDUCTING SUPER COLLIDER DIPOLE

Thomas H. Nicol

Fermi National Accelerator Laboratory

P.O. Box 500, Batavia, IL   60510, USA

INTRODUCTION

The cryostat of an SSC dipole magnet consists of all magnet components except the cold mass assembly. It serves to support the cold mass accurately and reliably within the vacuum vessel, provide all required cryogenic piping, and to insulate the cold mass from heat radiated and conducted from the environment. It must function reliably during storage, shipping and handling, normal magnet operation, quenches, and seismic excitations, and must be manufacturable at low cost.

The major components of the cryostat are the vacuum vessel, thermal shields, multilayer insulation (MLI) system, cryogenic piping, interconnections, and suspension system. Figure 1 is a cross section through an iteration B SSC dipole magnet and illustrates the major cryostat components.

The overall cryostat design required consideration of fluid flow, proper selection of materials for their thermal and structural performance at both ambient and operating temperature, and knowledge of the environment to which the magnets would be subjected over the course their twenty year expected life. Analytical models were constructed where appropriate and tested whenever possible. The design effort has involved close interaction of magnet, cryogenic system, and manufacturing designers.

It is safe to say that the cryostat design was driven to a large extent by the heat load budget. Table 1 lists the allowable heat load to each thermal station for later reference.1

VACUUM VESSEL

The vacuum vessel provides the insulating vacuum required to minimize the effects of heat transfer to the cold mass by residual gas conduction and serves to transfer static and dy-

New Techniques for Future Accelerators III
Edited by G. Torelli, Plenum Press, New York, 1990

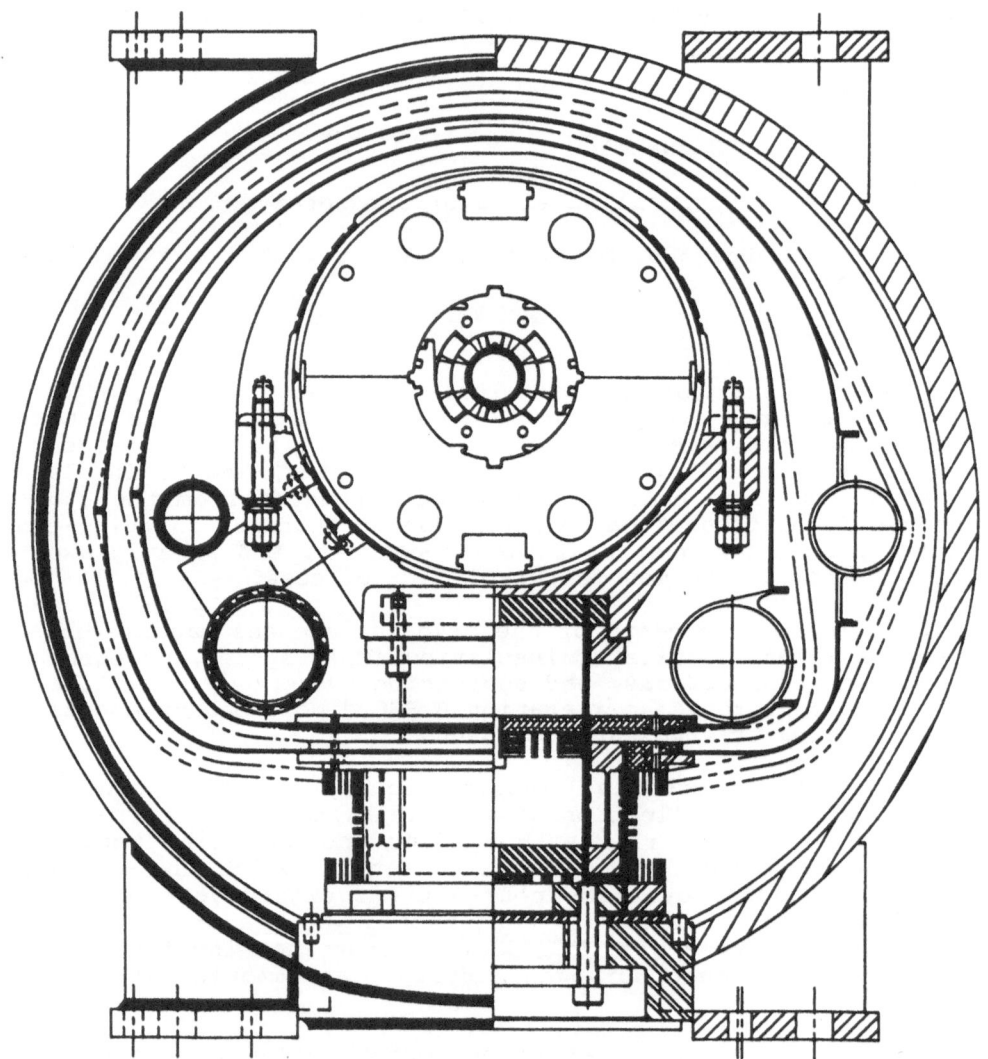

Fig.1. SSC Dipole Cryostat Cross Section

Table 1. Dipole Heat Leak Budget (Watts)

Circuit	4.5K	20K	80K
Static			
Infrared	0.05	2.16	17.7
Supports	0.12	0.82	7.2
Connection/instr	0.15	0.32	2.1
Subtotal	0.32	3.30	27.0
Beam related			
Synch radiation	2.34		
Splices	0.10		
Beam gas	0.10		
Beam microwave	0.20		
Subtotal	2.74		
Dipole Totals	3.06	3.30	27.0

namic loads to ground. It is circular in cross section, has an interconnection bellows at one end, and is connected to ground by means of support assemblies.

The vessel is fabricated from six sections of 609.6 mm OD, 6.35 mm wall SA516 steel pipe. These sections are joined at reinforcing castings at each of five internal support locations. The reinforced sections are required to transfer suspension loads to the vessel without imposing high bending stresses in the vacuum vessel walls. The reinforcing castings are 19.05 mm thick. The SA516 steel offers greater ductility at low temperature than does mild steel, but represents only a minor cost penalty.

Two of the reinforcing rings are at the external support locations. They support the weight of the magnet assembly and serve as the attachment points for the magnet to the accelerator tunnel floor. This configuration requires that the vacuum vessel be pre-curved at assembly to offset the effects of static sag in the vacuum vessel due to the weight of the cold mass. The required curve is approximately 9.53 mm from center to end.

If this pre-curving scheme proves impractical the external foot locations can be moved to a more optimal position with respect to vacuum vessel deflections. Locating the external supports at +/- 4.57 m eliminates the need for precurving because it minimizes the vacuum vessel sag. The drawback is that it requires seven reinforcing rings rather than five, resulting in an increase in the number of parts required and consequently an increased cost.

The completed internal cryostat assembly is inserted into the vacuum vessel on a tow tray installed at the bottom of the vessel shell. Figure 2 shows a layout view of the vacuum vessel assembly.

THERMAL SHIELDS

The cryostat has two thermal shields which intercept heat radiated and conducted from the environment. The shields enti-

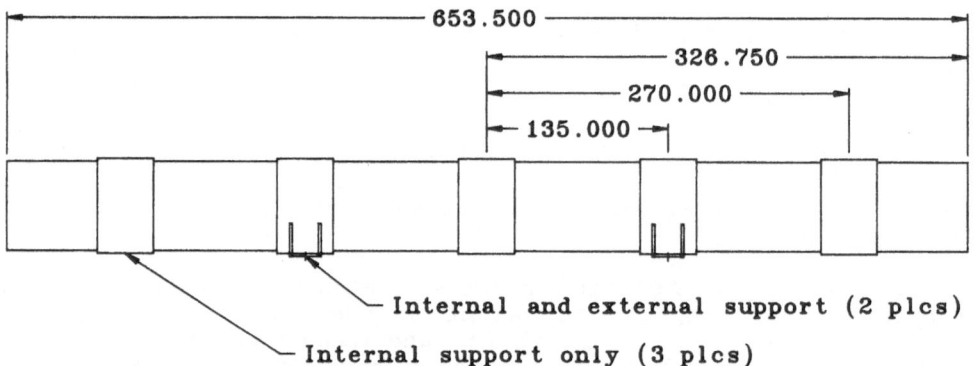

Fig.2. SSC Dipole Cryostat Vacuum Vessel

rely surround the cold mass assembly. The outer shield is cooled by liquid nitrogen and operates at 80K. The inner shield is cooled by gaseous helium and is controlled to operate at approximately 20K.

Each shield is fabricated from three pieces. Top and bottom sections are formed from 1.59 mm thick 6061-T6 aluminum which has high thermal conductivity and remains ductile at cryogenic temperatures. The cooling pipes are 6063 aluminum extrusions. The shield sections are welded to each other and to the extruded pipes and effect both structural and thermal bonds.

The shields are supported vertically and laterally at five points at the support posts. They are allowed to slide axially with respect to the support posts at all but the center location to accommodate axial shrinkage during cooldown. Attachment to each post is made by lifting the shield bottom up to the post intercept ring and sliding a retainer under the shield into a slot in the intercept. The shields are thermally anchored to the support posts by copper cables crimped to lugs welded to the shields and the post intercepts prior to assembly.

CRYOGENIC PIPING

The cryogenic system requires four pipes in the cryostat exclusive of the cold mass containment vessel. The pipes provide services for the 80K liquid nitrogen shield, 20K gaseous helium shield, 4.5K liquid helium return, and 4.5K gaseous helium return. The liquid and gaseous helium returns pipes are 316L stainless steel. They are supported by hangers welded to the cold mass, and are thermally insulated from the cold mass by G-10 bushings. The 80K and 20K shield pipes are 6063 aluminum extrusions. They are welded to their respective shields for both structural integrity and good thermal contact.

All of the pipes must be welded to stainless steel bellows in the interconnection area. The 80K and 20K shield pipes require the use of aluminum to stainless steel transition joints. Several commercially available transition joint types have been tested; diffusion bonded, explosion bonded, and brazed. Samples of each type have been subjected to repeated cold shock and vacuum leak check tests without a single failure. Table 2 lists the current effective size for each of the four cryostat pipes.

Table 2. Cryostat Pipe Sizes (mm)

Pipe	OD	ID
Liquid helium return	48.26	44.96
Gaseous helium return	88.90	84.68
80K Liquid nitrogen shield	69.85	63.50
20K Gaseous helium shield	82.55	76.20

MULTILAYER INSULATION

The insulation system limits heat leak from thermal radiation and residual gas conduction to the levels listed in Table 1.[2] It must perform reliably over the 20 year life of the SSC in an environment which includes irradiation, many thermal cycles, and periods of poor insulating vacuum. Essential to meeting these requirements is an insulation system design which addresses transient conditions through high layer density for gas conduction shielding and which has sufficient mass and heat capacity to lessen the effects of thermal transients.

The MLI system for SSC magnets consists of full length assemblies of aluminized polyester film fabricated in the form of blankets and installed on the 4.5K cold mass and the 20K and 80K thermal radiation shields. The thermal shield blankets are 17 m long and approximately 1.8 m wide. Thin layers of spun-bonded polyester sheet separate and space the reflective layers. Thicker layers of spunbonded polyester material cover the blanket top and bottom surfaces. Two MLI blankets are installed on the 80K shield, each blanket having 32 reflective layers. Two similar blankets, each containing 16 reflective layers, are installed on the 20K shield. The cold mass insulation consists of 10 reflective layers which are installed by wrapping an assembly of 5 reflective layers with spacers twice around the cold mass in a continuous wrap for a buildup of 10 layers.

After a careful study of the available literature, polyethylene terephthalate (PET) was selected as the best choice for each of the MLI blanket components. It provides the best combination of mechanical strength, radiation resistance, hygroscopicity, outgassing rate, and thermal performance of all commonly used materials. The blankets incorporate reflective layers of double aluminized PET film each separated by a single layer of spunbonded PET material. The thicker layer top and bottom layers of spunbonded PET material also position polyester hook and loop fasteners at the blanket edges. The fasteners are affixed to the cover layers by sewing. A thicker PET layer is also located midway in the blanket and separates the upper and lower 16 reflective layers. The multiple blanket layers are sewn together as an assembly along both edges of the blanket. Polyester thread is used in all sewing processes. At each blanket edge, the upper cover and MLI layers are sewn together through to the midlayer with the thread terminated in the thicker midlayer. The seam location is incremented laterally along the midlayer where the lower MLI layers are sewn through to the lower cover layer. The resulting step in the sewn seam serves to reduce heat conduction through the blanket by interrupting and lengthening the solid conduction path of the thread. Each blanket is fitted with openings through which the suspension system for the cold mass and shield assemblies protrude.

SUSPENSION SYSTEM

The cold mass assembly and thermal shields are supported by the suspension system.[3]. The suspension system serves to support the static weight of these components, to resist vertical, lateral, and axial loads resulting from shipping and handling, quench induced loads, and seismic excitations.

Table 3. Suspension Structural Constraints

Shipping and handling loads: 2.0 G (vertical)
1.0 G (lateral)

                                      1.5 G (axial)

Seismic load guidelines: Nuclear Regulatory
Guide 1.61, vertical
and horizontal spectra
scaled by 0.3

Maximum quench load: 111.2 kN (axial)

It must satisfy these mechanical constraints within the confines of a very stringent conduction heat load budget and must be high reliability, dimensionally stable, be easy to install and adjust, and have low cost.

The structural design constraints are listed in Table 3.[1] The thermal constraints are given in Table 1.

Support Post

The cold mass and shields are supported at five points along their length. The number and location of the support points was determined by analysis of the deflection of the cold mass assembly given a maximum allowable sag due to self weight of 0.25 mm. After consideration of tension member, compression member, elliptical arch, and post type supports, reentrant post type supports were selected.

The details of the reentrant post support are indicated in Figure 3. The post consists of two composite tubes and one stainless steel tube connected by metal discs and rings. The outer composite tube is filament wound using S-glass in an epoxy matrix. The inner tube is wound using a graphite fiber in an epoxy matrix. The composite tubes serve to carry all of the structural loads on the suspension system and to insulate the cold mass and shield assemblies from heat conducted from the environment. The metal discs and rings serve to joint the tubes and act as attachment points. The post assemblies use no chemical or mechanical fasteners to secure the metallic components to the composites. Rather all joints are effected using shrink fit techniques. That is, the composite tubes are sandwiched between the metal fittings with an interference fit, thereby securing the tube.

To ascertain the long term reliability of shrink fit joints, creep tests on a typical joint were made early in the development cycle. When extrapolated over 20 years these tests indicate no significant loss in joint strength. A post support is subject to thermal radiation that can significantly affect thermal performance. To minimize the effect of this radiation, multilayer insulation is used in the post assembly. Proper design of the thermal connections between the 20 K and 80 K intercepts of the supports and the thermal shields is also essential to minimize heat leak. Temperature rises of 5K and 1K are allowed at the 80K and 20K shield connections respectively.

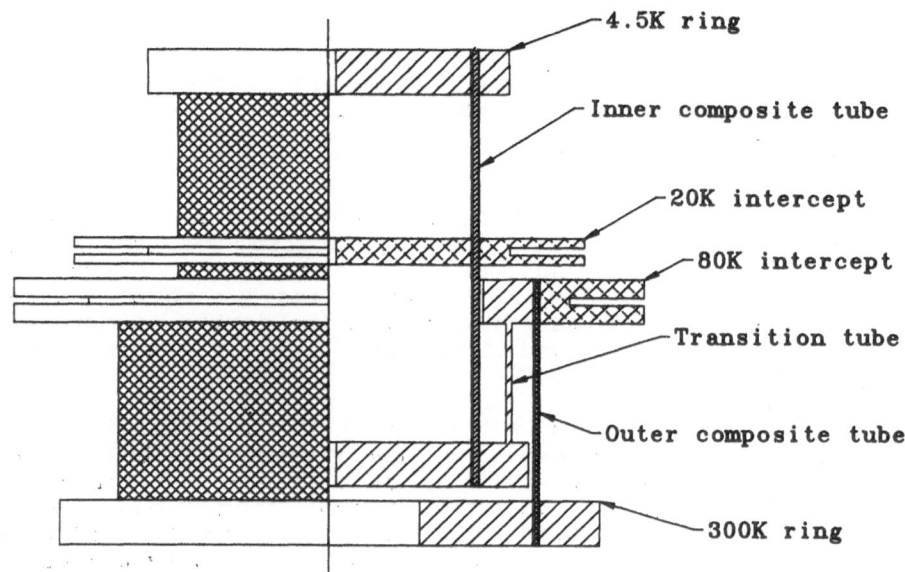

Fig.3. SSC Dipole Reentrant Support Post Cross Section

Post type supports performed as expected, both structurally and thermally, in several test assemblies in the R&D program. A reentrant tube support, instrumented with temperaure sensors, was installed and evaluated in a specially configured heat leak measurement dewar. Figure 4 illustrates a typical set of data resulting from the thermal performance tests of support post assemblies. The measured temperature profiles and heat leaks are in good agreement with predicted values. Performance in an actual SSC cryostat is expected to match this performance.

Anchor System

The anchor system serves to fix the axial and rotational position of the cold mass assembly relative to the cryostat vacuum vessel. It resists axial shipping and handling loads, quench loads and, axial components of seismic excitations. The anchor utilizes the collective bending strength of all five support posts to resist axial loads. Due to the fact that the cold mass slides with respect to the four posts outboard from center, the tops of all five supports are connected via graphite tie bars. Figure 5 illustrates the connectivity of all of the suspension components.

In order for the load distribution among all five supports to be ideal; i.e. each support post receiving 20% of an imposed axial load, the tie bars would need to be infinitely stiff. The actual load distribution (calculated) is given in Table 4.[3]

Table 4. Calculated Support Post Load Distribution

Center post:	26.4%
Posts at +/-3.43 m:	19.9%
Posts at +/-6.86 m:	16.9%

The measured load distribution from long magnet model DD0014 quench data is 23.1%, 18.3%, and 17.3% respectively.[4]

In addition to high stiffness, the tie bars must be dimensionally stable when cooled from room temperature to operating temperature (4.5K). The importance of this is due to the fact that one end of each support post is anchored to the vacuum vessel at 300K. Excessive shrinkage of the tie bars during cooldown would impose high bending stresses on the support post composite tubes. Graphite fibers tend to grow when cooled and the epoxy matrix of the composite shrinks. By tailoring the fiber volume, fiber type, and wind angle, a tie bar can be manufactured whose length remains virtually constant when cooled from 300K to 4.5K. Tests on prototype tie bars indicated a very small coefficient of thermal expansion. A tie bar 3.05 m long shrunk only 0.025 mm when cooled from 300K to 80K. The length change during cooldown to 4.5K has yet to be measured.

Cradles

The cold mass assembly is mounted on each of the five support posts with a cradle assembly. The cradle consists of

5x7 CgRP/GRP CCS WITH SLIDE – TEMPERATURE PROFILE MEASUREMENTS (K)

SENSOR NUMBER	SENSOR TYPE	DESIGN TEMP. (K)	PRIMARY INTERCEPT TEMPERATURE (K)							
			11.4	13.4	22.5	23.2	27.1	31.7	32.6	37.9
1	CARBON RES.	4.5	6.2	6.5	8.0	9.4	8.9	9.9	10.1	11.5
2	CARBON RES.	4.5	6.3	6.6	8.3	9.7	9.2	10.3	10.5	12.0
3	CARBON RES.	4.9	6.9	7.5	10.2	11.5	11.9	13.6	13.9	16.2
4	CARBON RES.	4.9	6.9	7.3	9.8	11.2	11.3	13.0	13.3	15.3
5	CARBON RES.	13.6	9.5	10.7	16.8	17.7	19.9	23.5	24.2	28.2
6	CARBON RES.	13.6	9.3	10.4	16.4	17.3	19.5	22.8	23.4	27.6
7	CARBON RES.	20.0	11.9	13.5	21.9	22.6	26.2	30.5	31.3	36.5
8	CARBON RES.	20.0	12.2	13.9	22.2	22.9	26.7	31.2	32.2	38.2
9	CARBON RES.	20.0	15.9	17.4	25.0	25.6	28.9	33.2	34.0	38.9
10	CARBON RES.	20.0	11.4	13.4	22.5	23.2	27.1	31.7	32.6	37.9
11	CARBON RES.	20.0	30.9	31.7	35.8	36.2	39.2	40.8	41.4	44.6
12	CARBON RES.	20.0	OPEN	OPEN	OPEN	OPEN	OPEN	OPEN	OPEN	OPEN
13	Pt RTD	79.0	83.1	83.1	83.0	83.4	83.5	83.5	83.8	84.2
14	Pt RTD	79.1	91.2	91.1	90.9	91.2	91.2	91.1	91.4	91.5
15	Pt RTD	79.1	87.2	87.2	86.9	87.3	87.2	87.2	87.4	87.6
16	Pt RTD	80.0	84.9	84.8	84.5	84.9	84.8	84.6	84.9	85.0
17	Pt RTD	80.0	86.8	86.7	86.4	86.7	86.7	86.5	86.8	86.9
18	Pt RTD	80.8	110.3	110.2	109.9	110.2	110.2	110.1	110.3	110.4
19	Pt RTD	200.0	204.0	203.8	203.6	203.8	203.9	204.1	204.0	204.2
20	Pt RTD	200.0	201.1	201.0	200.9	201.1	201.2	201.4	201.3	201.5
21	Pt RTD	299.7	267.7	267.3	267.2	267.5	267.6	268.0	267.7	267.9
22	Pt RTD	300.0	281.9	281.6	281.5	281.8	282.0	282.3	282.0	282.3
23	Pt RTD	300.0	282.6	282.3	282.5	282.5	282.6	282.3	282.7	283.0

HEAT LEAK MEASUREMENTS (mW) BY HEATMETER & MATERIAL PROPERTIES

	13	15	30	CALIB.	41	56	58	80
Q_{HM}	13	13	25	24	35	50	53	75
Q_2	7	9	9	24	35	50	53	75
Q_{10}	466	460	429	431	412	383	381	341
Q_{16}	1566	1566	1583	1582	1594	1613	1607	1628

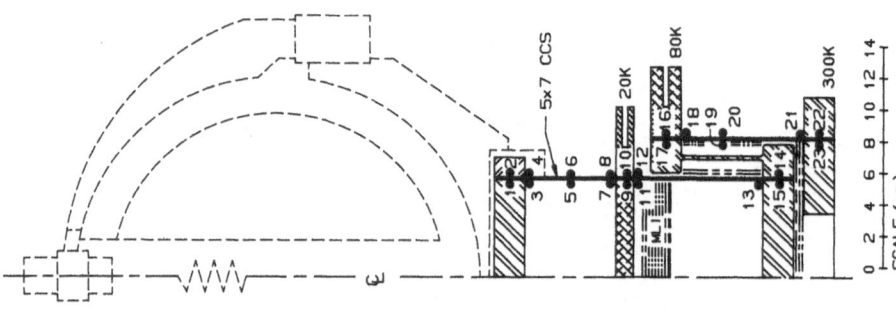

Fig. 4. SSC Dipole Support Post Test Data

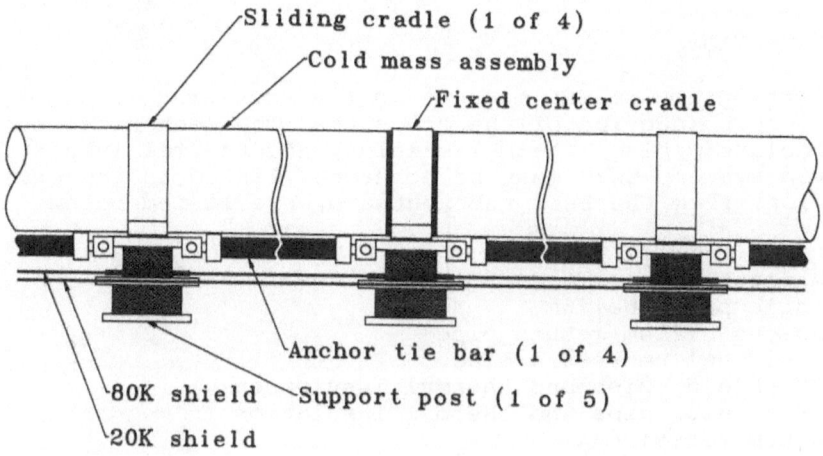

Fig.5. SSC Dipole Cryostat Suspension System Components

two pieces; a lower and an upper half. The lower half is connected to the 4.5K ring of the support post by a snubbing collar. The cold mass rests in the lower half. The top cradle half is bolted to the lower half. Conical spring washers act as the preload mechanism to allow for radial shrinkage of the cold mass during cooldown.

Each of the slide cradles contains four DU pads which serve as bearings. DU is a Teflon and lead impregnated, sintered bronze matrix on a steel backing plate. Tests of actual bearing assemblies indicate that at operating temperature and in vacuum, the coefficient of friction between the pad and cold mass is approximately 0.3.

The center anchor cradle is fixed axially and rotationally with respect to the cold mass. This is accomplished by welding a band to the cold mass and capturing this band inside a groove cut in the cradle.

INTERCONNECTIONS

Installation of SSC magnets in the accelerator tunnel requires interconnection of the cryostats. The mechanical connections include piping, thermal shields, and thermal insulation. The seven systems that must be connected, listed in the order of connection, from the beam tube outward, are listed below.

Beam tube
Single phase helium
Gaseous helium return pipe
Liquid helium return pipe
20K shield, pipe and thermal insulation
80K shield, pipe and thermal insulation
Vacuum vessel

All pipe connections are made by vacuum tight welds and meet all system pressure requirements. The welds are made by automatic computer-controlled welding units. When disconnection is necessary, the welds will by removed by orbital pipe cutters.

Each connection includes a bellows which provides enough longitudinal travel to compensate for magnet cooldown shrinkage; approximately 50 mm at the interconnect area. All bellows are made of 316L stainless steel.

The interconnection region is shielded and thermally insulated in the same manner as the body of the magnet. Shield bridges are installed which are similar in cross section to the body shields and similarly insulated. They are attached by a rigid mechanical connection to one magnet and overlapped but not attached to the other to allow for magnet to magnet contraction during cooldown. Copper cables provide the thermal connection between the magnet body shields and the shield bridges. Cables are welded to both the body shield and the shield bridge during magnet construction. They are then connected by crimping during the interconnection process. The shield bridges have openings which facilitate pumpdown and prevent large pressure differences from occurring between different areas in the cryostat.

The interconnection blankets are comparable in design to the cryostat body blankets. The junction of the cryostat and interconnection blankets is accomplished by means of over-lapping stepped ends at the blanket interfaces. Cryostat body blanket stepped ends nest with counter-stepped ends of the interconnection blankets. To assure that the mating steps move together during thermal cycles and do not separate, the blanket assemblies are installed such that the overlapped steps are carried between cover layers which bridge the interconnection region and which are firmly connected to corresponding cover layers on the cryostat blankets. To facilitate the intended performance under thermal excursions, a single sheet of cover layer material is first installed to the shield bridge. The cover layer material spans the shield bridge and overlaps its ends with the cryostat blanket cover layer that is against the shield. The cover layers are fastened to each other around the perimeter of the each overlapped end by means of staples or heat fusing. As the interconnection blankets are subsequently assembled around the shield bridge, the top cover layer of each interconnection blanket is similarly secured about its perimeter to the corresponding cover layer of the cryostat blankets. Cuts are made through the MLI following each individual blanket installation to access the 20K or 80K shield pressure vent located at the crown of each shield bridge.

SUMMARY

The SSC development program has afforded us the opportunity to extend the design of cryostats for superconducting magnets far beyond the state of the art present at the end of the Tevatron program. Advances in new materials technology have opened up options for cryostat designers in both thermal and structural materials. Strict limits on allowable heat load have forced us to develop new mechanisms for structural support and thermal shielding. The end result is a cryostat design which meets the demands of the SSC and which will serve as the starting point for the development of other magnet systems far into the future.

REFERENCES

1. SSC Central Design Group, "Superconducting Super Collider Magnet Systems Requirements," SSC-100, October 1986.
2. J.D. Gonczy et al, Thermal Performance of a 100 Percent Polyester MLI System for the Superconducting Super Collider; Part II: Laboratory Results (300K - 80K), presented at the CEC/ICMC-89, Los Angeles, CA, July, 1989.
3. T.H. Nicol et al, Design and Analysis of the SSC Dipole Magnet Suspension System, presented at the First International Symposium on the Super Collider, New Orleans, LA, February 1989.
4. T.H. Nicol, Structural Performance of the First SSC Design B Dipole Magnet, presented at the CEC/ICMC-89, Los Angeles, CA, July, 1989.

DIPOLE MAGNET DEVELOPMENT FOR THE RHIC ACCELERATOR

P.Wanderer, J.Cottingham, G.Ganetis, M.Garber,
A.Ghosh, C.Goodzeit, A.Greene, R.Gupta, J.Herrera,
S.Kahn, E.Kelly, G.Morgan, J.Muratore, A.Prodell,
M.Rehak, E.P.Rohrer, W.Sampson, R.Shutt, P.Thompson,
E.Willen

Brookhaven National Laboratory[*] , Upton, N.Y., 11973
U.S.A.

INTRODUCTION

A Relativistic Heavy Ion Collider is presently in the design stage at Brookhaven National Laboratory in New York. It will collide beams of nuclei as heavy as gold, accelerated in two storage rings to energies between 7 and 100 GeV/u. The conventional facilities and injectors for the collider are largely in place. These include the 3.8 km long tunnel enclosure (built as part of the CBA project), four of the six planned experimental halls, a large control center building, a 25 kW helium refrigerator, and injection tunnels from the AGS. (The AGS is presently engaged in a fixed-target program of heavy ion experiments. A Booster synchrotron under construction will allow extending the mass of ion species from sulfur to gold.)

The collider lattice reflects the need for strong focussing to maintain a small beam size while coping with the severe intrabeam scattering of heavy ion beams. It is based on one dipole per half cell. The regular arcs comprise 72 cells per ring, requiring a total RHIC arc magnet inventory of 288 dipoles and 276 quadrupoles. Adjoining each quadrupole will be a separate sextupole as well as a multipole corrector. In addition, there is a large complement of standard and large-aperture magnets associated with the interaction regions.

Magnet R and D is well underway[1]. Eight full-length dipoles and two of each of the other magnets in a half cell (quadrupoles, sextupoles, and correction magnets) have been built. For this Workshop, I will report on our work on the dipole because it is somewhat different from current highfield designs and because it is the largest single cost item in the accelerator.

---

[*]Work supported by U.S. Department of Energy

A heavy-ion machine in the relativistic energy regime was an option in the original CBA design. With many completed facilities (refrigerator, tunnel, magnet development, etc.) it was realized that an attractive physics program could be realized at much reduced cost with a dedicated heavy ion collider. With a conservative magnet design based on CBA experience and improvements in superconductor current-carrying capacity, it was felt that the energies achievable should be sufficient for exploring the regime of quark-gluon plasmas, a topic of interest both to nuclear and elementary particle physicists. The design of RHIC has been optimized for heavy ions (e.g., more focussing, large aperture), but the machine can also be operated as a proton-proton collider at luminosities in the range 10^{31} to 10^{32}.

From these considerations, the design guidelines for the dipole magnets were: curved magnets to maximize use of magnet aperture, no distributed correction elements, 30% design central field margin for heavy ion operation at 3.5 T, 10% margin for proton operation at 4.2 T, mechanical design capable of achieving 5 T operation, single layer coil, cold iron design, SSC-dimensioned cable. The RHIC magnets were designed to make optimal use of existing technology with a generous margin so that a quick start could be made. This is in contrast to the SSC, where the larger cost of the magnet system dictated the need for a more carefully optimized design with a resulting longer development time.

DIPOLE MAGNET DESIGN

The magnet cold mass (Fig.1) has a relatively large bore

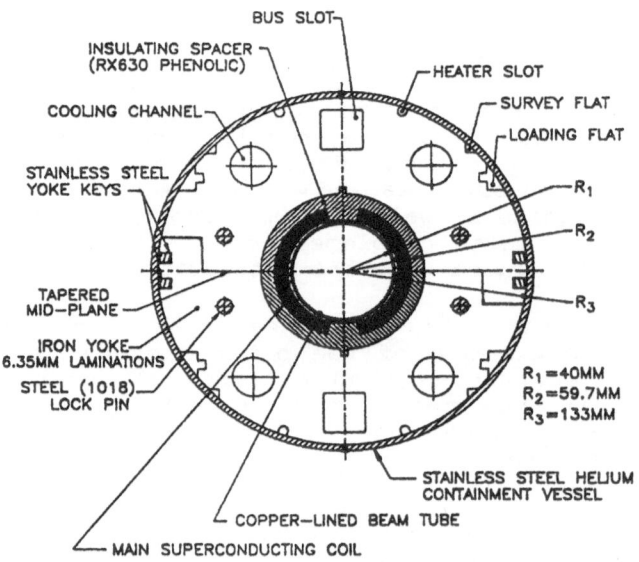

Fig. 1. Cross section of the RHIC design D dipole.

(80 mm) to accommodate the emittance growth associated with intrabeam scattering, a modest operating field (3.45 T), a single layer cosine theta coil, an iron yoke assembled as collars to prestress the coil, and no internal trim coils. The effective magnetic length is 9.46 m. Dipole construction and operation parameters are listed in Tables I and II.

Table I. Dipole Construction Parameters

Beam tube, inner radius	36.45 mm
Outer radius	38.10 mm
Coil, number of turns	32
Inner radius (R_1)	39.9 mm
Outer radius	50.0 mm
Overall length	9.65 m
Iron yoke, inner radius (R_2)	59.69 mm
Outer radius (R_3)	133.35 mm
Length (iron)	9.64 m
Length, including end plates	9.71 m
Radius of curvature, nominal	243 m
Sagitta	46 mm
Weight including coil	3605 kg
Assembly prestress	
Room temperature	10 kpsi
Cold (after relaxation)	> 4.8 kpsi

Table II. Dipole Operating Parameters

Effective length	9.46 m
Inductance	28 mH
Field at injection	0.4 T

At 100 GeV/u Operation

Current	4.98 kA
Field	3.45 T
Ramp rate (60 sec)	0.05 T/sec
	73 A/sec
Stored energy	351 kJ
Stored energy/sextant	8.42 MJ

At Predicted Quench (4.6 K)

Current	6.6 kA
Field	4.4 T
Stored energy	589 kJ
Current density in copper	0.96 kA/mm^2

To minimize cost and take advantage of existing technology, the magnet uses superconducting cable of the same type as the cable developed for the outer layer of the SSC dipole. The specifications for the wire and the cable are given in Table III. A cabling degradation of 10% is assumed. With this choice of cable, the ratio of quench current to operating current at 100 GeV/u is 1.3. A cross section of the cable is shown in Fig. 2. The cable has the same dimensions as the SSC cable, but a higher value for the copper-to-superconductor ratio (2.25 vs. 1.8). From experiments in the BNL cable test facility there is evidence that the RHIC dipole needs more copper

Table III. Parameters for Superconducting Wire and Cable

Wire		
Critical current*	>264	A
Critical current density in superconductor*	2600	A/mm^2
Nominal copper-to-superconductor ratio	(2.25±0.1):1	
Wire diameter	0.648±0.003	mm
Filament size	6	μm
Filament spacing	>1	μm
Number of filaments	3600	
Wire twist pitch	12.7±1.3	mm
Wire twist direction	right	
Maximum resistance at 295 K	750	μΩ/cm
Minimum RRR	90	
Critical current*	>7.13	kA
Number of wires	30	
Cable lay pitch	74±5	mm
Cable lay direction	left	
Keystone angle	1.2±0.1	deg
Cable width (bare)	9.73±0.03	mm
Mid-thickness (bare)	1.166±0.006	mm
Maximum resistance at 295 K	26.5	μΩ/cm
Minimum RRR	60	

*1 x 10^{-14} Ω·m @ 5 T and 4.2 K.

than the SSC dipole in order to be able to make full use of the current-carrying capacity of the NbTi. The added copper also makes quench protection easier. The cable is insulated with 0.025 mm-thick Kapton, overlapped to provide a double layer. Fiberglass tape, impregnated with B-stage epoxy, is wrapped over the Kapton. When cured, the epoxy and fiberglass provide the turn-to-turn bonding needed during the assembly process.

178

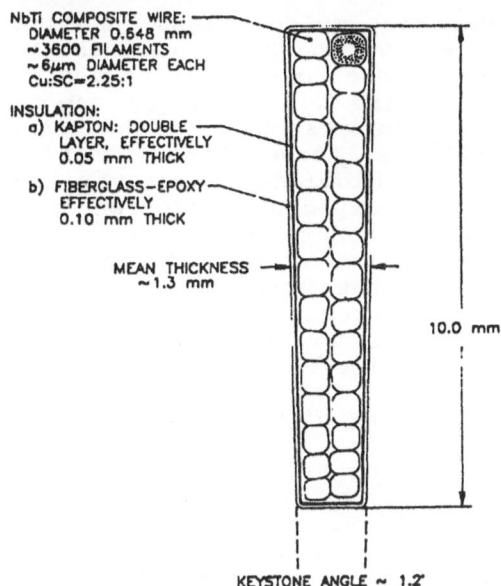

NbTi COMPOSITE WIRE:
DIAMETER 0.648 mm
~3600 FILAMENTS
~6μm DIAMETER EACH
Cu:SC≈2.25:1

INSULATION:
a) KAPTON: DOUBLE LAYER, EFFECTIVELY 0.05 mm THICK

b) FIBERGLASS—EPOXY EFFECTIVELY 0.10 mm THICK

MEAN THICKNESS ~1.3 mm

10.0 mm

KEYSTONE ANGLE ~ 1.2°

Fig.2 Cross section of superconducting cable

Because of concern about damaging the narrow edge of the cable, the conductor is rolled to a smaller keystone angle (1.2° rather than 1.7°) than that required for alignment along a radius. However, the centers of each of the five conductor blocks in the coil are positioned along a radius. The copper wedges provide the compensation angles to make this possible. By adjusting the sizes and positions of the four wedges and the coil pole spacer, it has been possible to design a single-layer coil with acceptable field harmonics. Optimization of the available parameters was done using harmonic requirements up to and including the 26-pole. Other criteria were also used in selecting the best design; these were:

1) three turns in the conductor block nearest the pole for enhanced quench propagation,
2) maximum transfer function,
3) reduced sensitivity to random variations in coil construction
4) reduced sensitivity to changes in cable thickness, and
5) design flexibility in case small changes in harmonics are required.

The coil ends which are located within the iron yoke were designed to minimize field harmonics and also to reduce field enhancements so that the ends do not provide an operational (quench) limit. This latter goal was achieved by allowing wide spacing of the turns nearest the pole spacer. Multiple, thin G-10 spacers coated with dry but uncured epoxy are used between turns. This makes it possible to wind the coil with flexible spacers which then become solid when the epoxy is cured. Therefore the coil end can withstand the axial Lorentz force (approximately 17,000 lbs. at peak field) when it bears against the end plate.

Prestress is applied to the coil directly by the iron

yoke through 10 mm thick molded glass-phenolic spacer-insulators surrounding the coil and keyed into the yoke laminations. Axial conductor motion due to end forces that are present during magnet operation are minimized by one-piece end plates 31.75 mm thick. During assembly, each end of the coil is preloaded with 1,500 lbs. force. A split stainless steel shell welded at the vertical midplane takes over the compression of the yoke when it is at cryogenic temperature. The dipoles are assembled in fixtures that introduce the required 47 mm sagitta. The sagitta is locked in place when the shell is welded. The shell and the bore tube are part of the helium pressure vessel. For increased reliability, the magnet system was designed so that it would not be necessary to use trim coils internal to the dipole. A lumped correction system outside the magnets corrects for harmonics at injection as well as for iron saturation at the highest operating field.

The cryostat housing the cold mass consists of a carbon steel vacuum vessel (610 mm outer diameter), an aluminum heat shield maintained at 55 K, blankets of multilayer aluminized Mylar insulation, cryogenic headers, and the magnet support system (Fig.3). The cold mass is supported at three locations by a folded, insulated post-type support similar to that developed at Fermilab for the SSC magnets.

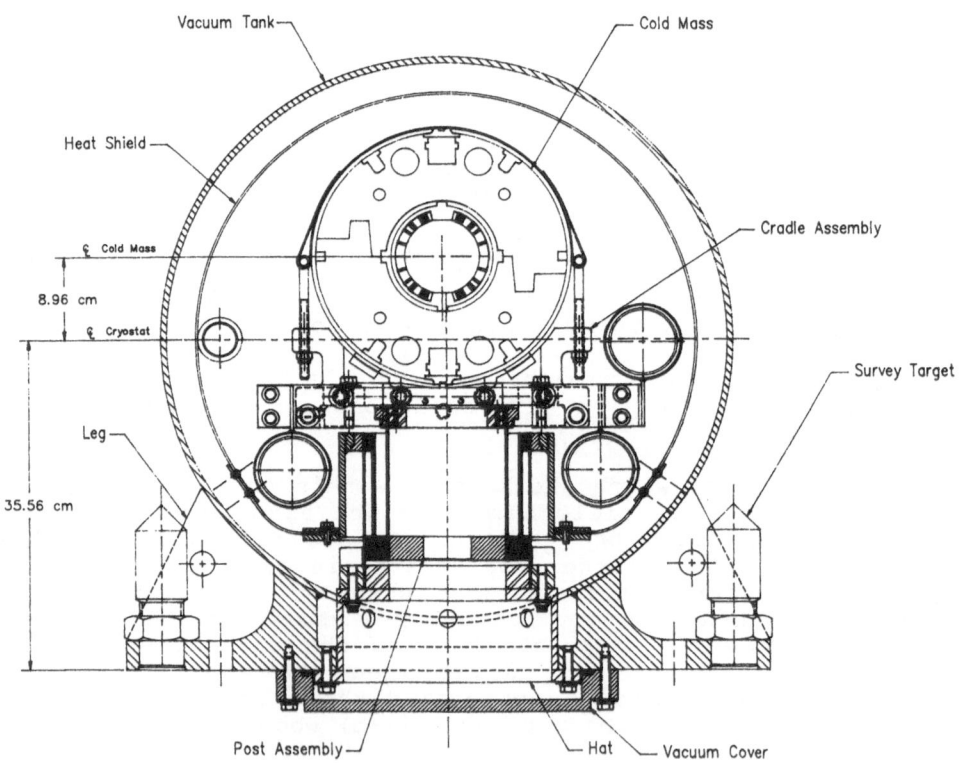

Fig.3. Cross section of cold mass and cryostat assembly showing survey target positions.

Training History

Of the eight full-length dipoles, six have been tested. (The remaining two will be tested in the near future.) All quenches of these first six magnets were above the RHIC operating current, but some training was observed. To gain information on how to reduce the training, strain gauge instrumentation developed initially as part of the SSC program was installed in the sixth magnet. With the help of the data from the strain gauges, it was possible to assemble the magnet with coil prestress much closer to the design value than for previous magnets. The training history of this magnet is shown in Fig.4. After one training quench, the magnet was nearly at the limit of the conductor. However, after thermal cycles, retraining and some fluctuations in quench current were observed.

The voltage tap instrumentation installed in the magnet allowed the location of these quenches to be identified as the turn next to the pole in the straight section of the magnet (four quenches) and the extension of this turn to the splice between upper and lower coil halves (the two quenches denoted R). After the assembly of the magnet it was discovered that the key

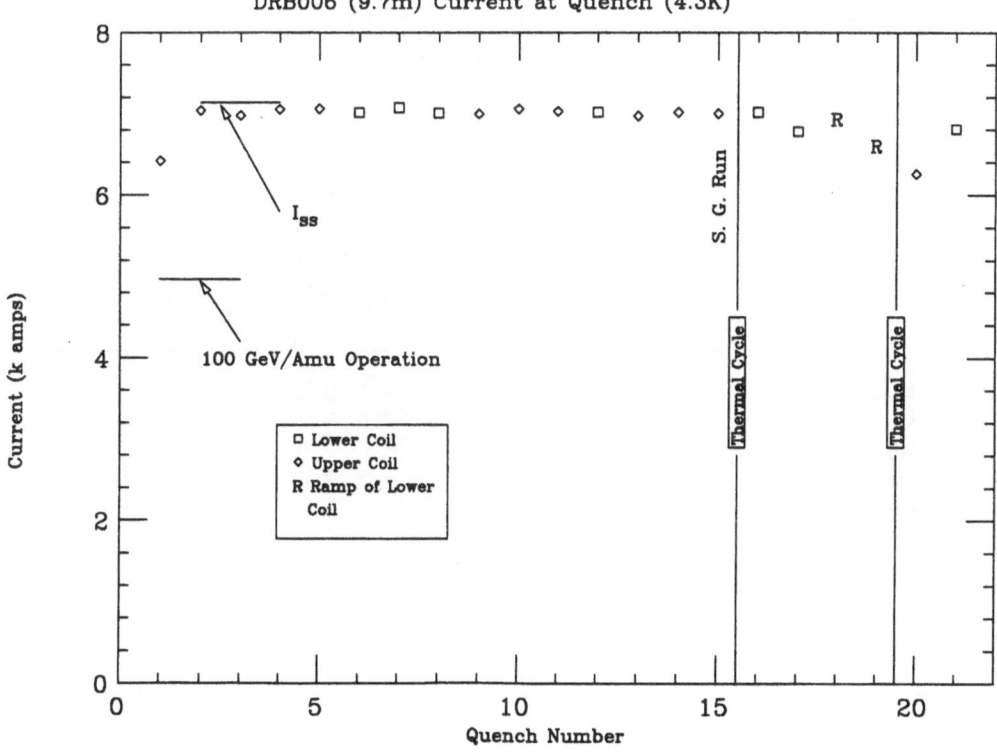

Fig.4.Quench test history of the most recent RHIC dipole. The ramp rate was 16 A/sec., except for the strain gauge run.

of the insulator, located at 90°, was slightly larger radially than the keyway in the yoke. As a result the turns adjacent to the pole lacked the support provided by the rigid yoke. These problems were corrected for the assembly of the two most recent magnets and improved performance is expected.

Field Quality

The field quality of accelerator magnets is characterized by harmonics which are defined in terms of the horizontal and vertical field components on the median plane:

$$B_y(X) = B_0 (b_0 + b_1 s + b_2 s^2 + \ldots)$$
$$B_x(X) = B_0 (a_0 + a_1 s + a_2 s^2 + \ldots)$$

where x is the transverse offset from the magnetic center, $s = x/(2.5 \text{ cm})$ is the offset normalized to the RHIC reference radius, and B_0 is the dipole field strength.

The variation of the allowed multipoles (b_n with n even) with current due to saturation and magnetization effects are important considerations in the magnet design. As a result of accelerator physics studies made after the design of the first four full-length dipoles the allowed saturation variation of the decapole term was reduced. To accomplish this, the distance between the coil and yoke was increased from 5 mm to 10 mm and a new coil cross section was designed. A short (1.8 m) model was built and tested to verify the new design. Measured values of the transfer function B/I, the sextupole b_2, and the decapole b_4 versus current are shown in Figs.5-7, respectively. The saturation variation of b_2, b_4, and b_6 satisfies the accelerator physics requirements[2].

Table IV. Corrected geometric multipole coefficients and estimated rms variations

	DRB5	DRB6	σ's
b_1	− .2	− 1.0	2.1
b_2	6.6	5.3	4.6
b_3	0	− .4	1.3
b_4	− 2.3	− 3.2	2.2
b_5	.1	.1	.5
b_6	0	.4	.8
a_1	3.5	1.3	4.3
a_2	.6	− .2	1.3
a_3	.3	.6	2.2
a_4	0	.2	.6
a_5	.6	.2	.9
a_6	0	0	.2

Values of the "geometric" multipole coefficients, those due to the coil cross section and the profile of the iron aperture, for the two most recent 9.7 m magnets are given in Table IV[3]. The measurements have been integrated over the straight section of the magnets. The values of the allowed terms are, of course, controlled in the coil design. Experience with assembly will probably allow the average values of the quadrupole terms (which are strongly affected by top-bottom and left-right asymmetries) to be kept near zero. The values of unallowed terms higher than the quadrupole are less than one unit even in these magnets. The multipoles of the magnet ends are also small.

In Table IV, the multipole coefficients are also compared to estimates of the rms variations for these magnets made on the basis of CBA and Tevatron magnet data. The procedure used is the same as that used to estimate errors for the SSC dipoles[4]. Roughly speaking, the multipole variations correspond to those expected from mechanical variations of order 0.05 mm. These estimates of rms variations are acceptable from accelerator physics considerations and serve as the benchmark for dipole field quality. Insofar as can be determined from two magnets, the magnet-to magnet variation is much less than the estimates.

For the two design B magnets, the average value of the transfer function in the magnet straight section, as determined from NMR measurements in 2.54 cm steps at 1.7 T, were 7.075 and 7.086 kG/kA. The transfer function varied by about 1×10^{-3} along the length of each magnet. These differences are expected to become smaller as more magnets are produced. The design and measured values of the effective length are close, 9.46 m and 9.43 m respectively. (Only one of the two magnets had an iron yoke which covered the coil ends.)

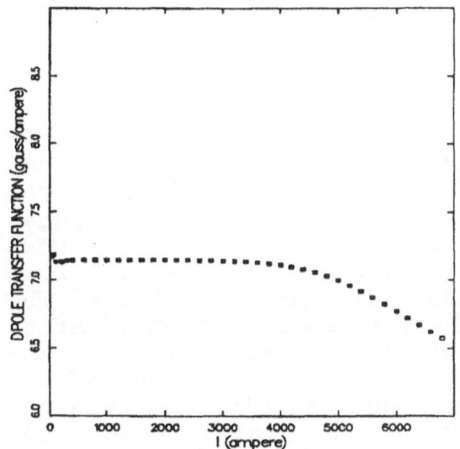

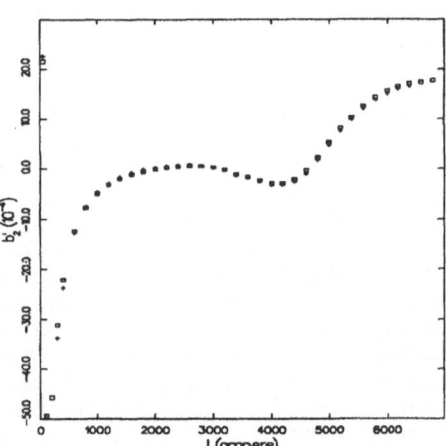

Fig.5. Measured transfer function (B/I) vs. current. The 100 GeV/u operating current is 5 kA.

Fig.6. Measured sextupole field vs. current. The field is evaluated at a radius of 2.5 cm, in units of 10^{-4} of the dipole.

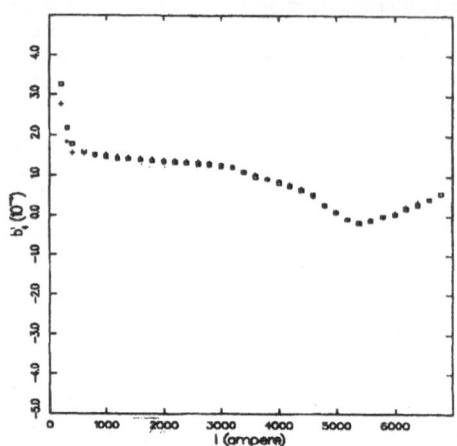

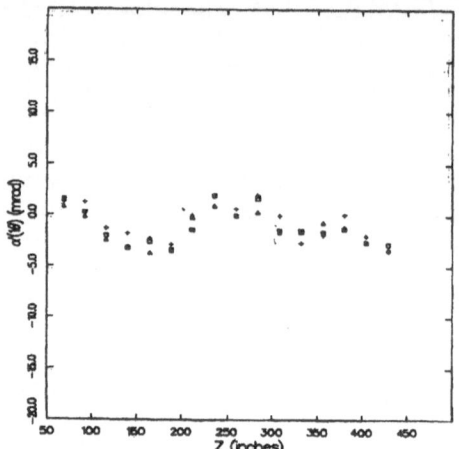

Fig.7. Measured decapole
    field vs. current.

Fig.8 Dipole field angle
    (mrad) along magnet
    axis at three different
    currents.

The two design B magnets were the first to use the post system of supports. At 1.2 kA, their average dipole angles were −1.5 mrad and −0.8 mrad. The axial variation in angle for one of the magnets is shown in Fig.8. The other magnet has a similar variation. In one of the magnets, the difference between the dipole angle at room temperature and at 1.2 kA was less than 0.1 mrad, but in the other it was about 2 mrad. Thus, the initial results are encouraging but some additional work will be needed to meet the 1 mrad RHIC rms tolerance on dipole angle.

Mechanical Behavior

The axial and azimuthal responses of the magnet to assembly, cooldown, and test were recorded with strain gauges[5]. The history of the magnet assembly as indicated by azimuthal gauges is shown in Fig.9. The lefthand portion of the abscissa shows the increase in coil stress as the pressure in the collaring press is increased and the keys installed on the collars. The coil stress drops about 5 kpsi when the press is removed after installation of the keys. The right-hand portion of the abscissa records the small loss of prestress due to creep and relaxation during the two months following assembly, and then the prestress at 4.5 K prior to excitation. The prestress loss due to creep and cooldown, averaged over the four strain gauges, was 2.3 kpsi.

After a number of quenches, the strain gauges were recorded as a function of current. A first-order approximation is that the forces should vary with the square of the current.

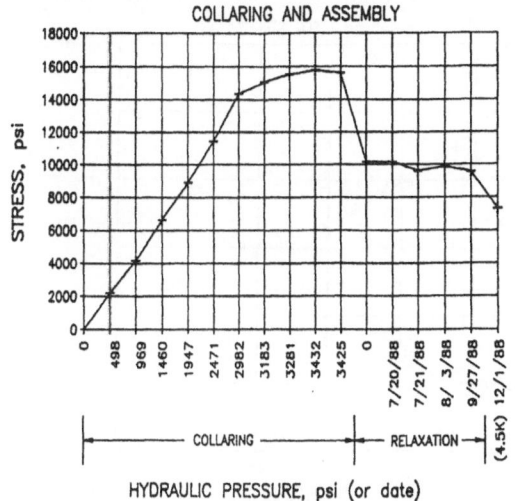

Fig.9. Assembly and cooldown history of coil stress in the most recent 9.7 m dipole, DRB006.

Coil stress and end force are plotted against current squared in Figs.10 and 11. The 100 GeV/u operating current is 5 kA, or 2.5×10^7 A^2 on the abscissa of these two plots. It is important that the coil be under azimuthal stress at all operating currents. In Fig.10 it can be seen that the coil stress decreases as the current is increased, indicating that the coil is in compression, as desired. The curve is close to linear. In Fig.11, the increase of pressure on the end is linear with the square of the current. The change of pressure with current is 2,800 pounds, much less than the 17,000 pounds of Lorentz force on the end at 5 kA. SSC magnet data are similar. The difference is due to friction between the coil and the insulator and yoke.

Quench Protection

Small heaters were placed on pole and midplane turns in the straight section of the coil so that quenches could be initiated at controlled positions and currents to study quench propagation and magnet selfprotection. The temperature at the spot heater location was determined from resistance measurements of the cable near the heater, with a small correction for the temperature distribution within this section of conductor. The temperature was correlated with $\int I^2 \, dt$. A convenient unit for this quantity is MIITS, defined as 10^6 A^2 seconds. For the heaters at the pole and on the midplane, the temperature was mapped as a function of current in 0.5 kA steps. Quenches initiated in the pole turn had higher temperatures than did quenches started at the midplane. The highest temperature, 400 K, was reached when the quench current was 4.5 kA, with $\int I^2 \, dt$ = 12.4 MIITS . This is substantially lower than the threshold for magnet damage, 835 K (13.8 MIITS). The temperature varied little within a 0.5 kA current range.

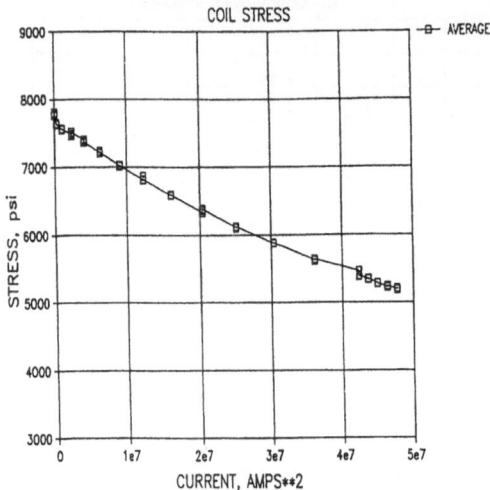

Fig. 10. Variation of azimuthal coil stress in magnet DRB006 as a function of the square of the current.

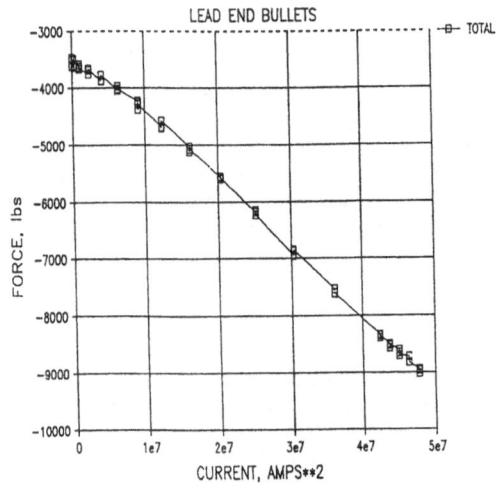

Fig. 11. Variation of axial force at the end of magnet DRB006 end as a function of the square of the current.

Individual magnets will be protected with a passive quench protection system, using diodes to bypass the current around the quenching magnet. One diode per magnet half will be satisfactory.

REFERENCES

1   P. Dahl et al., IEEE Trans. Magnetics 24, No. 2 (March 1988), p.723 summarizes earlier work on the RHIC dipoles.
2   "Conceptual Design of the Relativistic Heavy Ion Collider," Brookhaven National Laboratory Report 52195, May 1989 (unpublished).
3   To obtain the geometric terms, measurements were made with both positive and negative dI/dt at 1.2 kA where saturation effects are negligible. Special runs were made at currents where the sextupole and decapole were small enough that feeddown effects to quadrupole and octupole harmonics could be neglected. The allowed harmonics have been corrected for differences between the design sizes and actual sizes of the shims between the pole and the coil.
[4] J. Herrera et al., IEEE Trans. Nuclear Science, Vol. NS-32, No. 5 (October 1985), p. 3689.
[5] C. L. Goodzeit at al., IEEE Trans. Magnetics 25, No. 2 (March 1989), p. 1463.

FINE FILAMENT NbTi CONDUCTORS: DESIGN AND LARGE SCALE

PRODUCTION

M. Thöner

Vacuumschmelze GmbH

Gruner Weg 37, D 6450 Hanau F. R. Germany

According to the present specifications, the accelerators and storage rings of the next generations like LHC, SSC, UNK, RHIC and ELOISATRON will require beam line magnets wound from keystoned cables consisting of strands with fine filaments. The specified filament diameters are between 6 µm down to about 2.5 µm in order to reduce field distortions due to persistent currents inside the filaments. In addition, the filament spacing must be large enough to avoid coupling between the filaments. The required distance depends on the matrix material and a safe choice is of the order of 1 µm for pure Cu. It can be made much smaller for Cu-alloys like CuNi and CuMn (in case of NbTi) or CuSn (for Nb_3Sn).

Nb_3Sn conductors can be designed to meet these specifications and first prototype dipoles achieving the goal of reaching 10 T were tested recently[1]. Nevertheless the beam line magnets of the next generation with a field strength between 5 and 10 T most probably will be built with NbTi conductors because of the easier magnet technology, so that the following discussion concentrates on NbTi. For the sake of this short contribution it seems to be sufficient to translate the different specifications into a common specification at 5 T and 4.2 K. Typically superconductor critical current densities j_c of about 2500 A/mm^2 up to 3000 A/mm^2 in multifilamentary wires with 5 to 6 µm thick filaments will be needed. There exists the option to go down to 2.5 to 3 µm in case j_c-values will not degrade and the overall accelerator design can be made more cost effective. The level of critical current density given above is state of the art since several years now. NbTi conductors with this performance are produced on a large scale for industrial applications like magnetic resonance imaging (MRI) and high resolution spectroscopy (NMR) and also for the HERA dipoles and quadrupoles, indicating that there exist no intrinsic limitations in achieving these j_c values even on a large scale. Nevertheless practical limitations may be encountered mainly due to two reasons:

(1) Amount of available cold work especially at wire diameters above 1 mm

(2) Distortions of the filament uniformity along its length leading to a decrease of the effective cross section and therefore also of the current carrying capacity especially at filament diameters well below 10 μm.

The first effect is related to the fact that the optimization of j_C is achieved by creating a flux pinning effective microstructure consisting of normal conducting ribbon-like shaped α-Ti particles dispersed in the superconducting β-phase of NbTi. The optimun amount, size and shape of pinning centers is produced by a combination of cold working with intermediate heat treatments in the two phase region of the NbTi-phase diagram between 380°C and 420°C over many hours. To achieve above j_C values typically at least 3 heat treatments are required with sufficient initial, intermediate and final cold work, such that a certain minimum amount of area reduction is required. Record j_C values far beyond 3000 A/mm^2 can be reached by having a very large area reduction by cold work with many intermediate heat treatments. In contrast to that industrial processes include an extrusion step to allow for production of large quantities. The extrusion is done at elevated temperature to provide metallurgical bonding within the composite which in turn is needed to guarantee good workability with uniform area reduction of all components and to produce long piece lengths. In some kind of trade-off the extrusion parameters have to be carefully adjusted to simultaneously achieve good bonding and to reduce the "loss of memory" with respect to previous cold work. Typical industrial extrusion presses have billet diameters of 200 to 300 mm. As a consequence it is more difficult to achieve high j_C in wires with much larger diameters than 1 mm as compared with wire diameters between 0.5 to 0.9 mm.

There are several factors which can affect the uniformity of a filament and thereby of its current carrying capacity:

a) inhomogeneity of the NbTi alloy leading to nonuniformy pinning strength and/or deformation during cold work

b) formation of hard Cu-Ti-intermetallic compound particles at the interface to the matrix and subsequent nonuniform deformation

c) disturbance of the uniformity of the filament array due to stacking faults and/or insufficient filling factor in the billet

d) Filament sausaging due to the mismatch of the mechanical hardness of NbTi and the matrix.

The possible solutions for all four problem areas were identified already many years ago and their effectiveness was verified during the recent world wide conductor development efforts. Sufficiently homogeneous NbTi alloy is available and diffusion barriers e.g. of Nb are used to prevent the formation of Cu-Ti-intermetallics. Stacking techniques for very high filament numbers with high packing factors e. g. with hexagonal elements were developed. Sausaging can be avoided by selecting a small enough Cu:Sc-ratio in the multifilamentary area.

Obviously, several trade-offs have to be made. As an example, the Nb barrier has to be chosen with such a thickness that it remains effective during the last optimization heat treatment. On the other hand it should be as thin as possible for cost reasons and in order not to reduce the overall critical current density. An example of a single stacked wire with 6000 filaments is shown in Fig.1. The critical current density is 2800 A/mm^2 at 5 T with a wire and filament diameter of 0.648 mm and 5 μm, respectively[2].

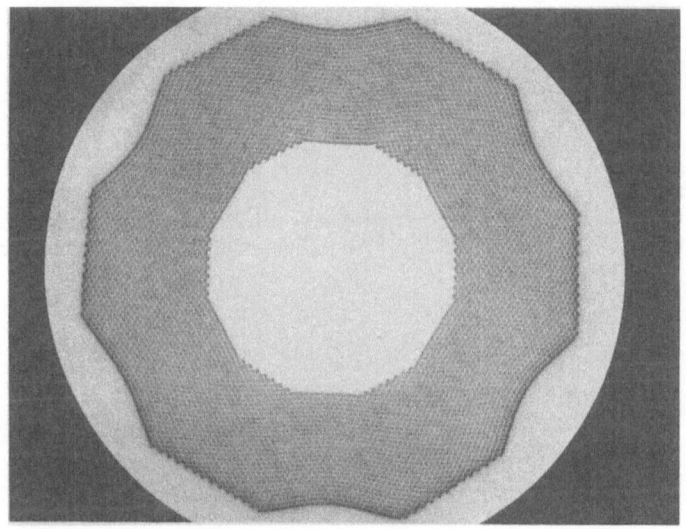

Fig.1. Cross section of a NbTi/Cu multifilamentary wire with 6000 filaments of 5 μm diameter (wire diameter 0.648 mm, Cu:SC-ratio = 1.8, critical current density 3000 A/mm^2 at 5 T and 4.2 K)

When using a pure Cu-matrix in the filamentary area 5 μm seems the lowest practicable filament diameter. At lower values the two contradictory requirements of avoiding filament coupling and preventing sausaging seem not to be reconcileable. Therefore a matrix material like CuNi or CuMn has to be chosen. At present CuMn is being preferred due to its expected smaller impact on conductor and magnet stability. The achievement of uniform fine filaments with 2.5 μm diameters is facilitated by the better matching of mechanical properties in combination with the possibility of narrow filament spacing. Due to the very high number of filaments (up to 40000 and more) double stacking and extrusion is the preferred route of fabrication in this case.

From the above said it becomes clear that several options exist for the production of NbTi fine filament conductors. Although trial fabrications were very successful, none of the processes is proven on a large scale such that no firm data are available on their relative reliability, variability of product performance and last not least cost effectiveness. Further research and development and conductor testing and characterization has to be performed before a sound basis for the selection of the best manufacturing route can be established. In addition, the impact on magnet and accelerator performance and overall cost must be carefully looked at. In all steps a close cooperation between conductor manufacturers, testing laboratories and magnet and machine designers is required. As for large scale production the schedule for installing of sufficient production capacity, adequate quality assurance and test equipment must be established timely, including the training of the different staffs.

REFERENCES

1.   A. Asner et al., to be published in the Proc. of the 11th
      International Magnet Technology, Tsukuba, August 28-Sept.
      1, 1989
2   H. Krauth et al., submitted to "1990 International
      Industrial Symposium on the Super Collider", Miami,Fl,
      March 14-16, 1990

CONTRIBUTION TO THE "ROUND TABLE ON THE INDUSTRIAL INVOLVEMENT

IN SC MAGNETS PRODUCTION"

S. Wenger

ELIN ENERGIEANWENDUNG GmbH
Magnet Technology Dept.

A-8160 Weiz/Austria

Elin, founded in 1892, ranges among the largest industrial enterprises in Austria. ELIN supplies engineering as well
as high quality products both for electric power application
and for electric power generation. The products include machines and electrical equipment for hydroelectric power plants
and thermal power plants as well as all associated electronic
devices for process control and automation. Based on this background ELIN is also involved in the design and manufacturing of
electromagnets for high energy physics.

ELIN, sited in Weiz with headquarters in Vienna, designs
and manufactures magnets for several laboratories and various
applications. A recent success was the design and construction
of a 10 Tesla Nb_3Sn model magnet, developed and successfully
tested in a collaboration with CERN.

Working together with highly sophisticated customers like
universities, international research institutes or national
laboratories very often necessitates unconventional project
management techniques. Tenders are frequently based on technical concepts rather than ready designs.

Engineering of practicable solutions during the production process presents a special challenge to a project manager
who however in most cases has to face a fixed budget and a
tight schedule. Laboratories could drastically improve this
situation if they convene to adapt a contract policy corresponding to the commercial conditions each industrial firm has
to live with: In case of a fixed-price-contract industry usually works on the basis of a detailed technical specification
with essentially zero product development cost; in the case of
a development project, particularily when a "first of the kind"
product has to be built, the contract should be on a cost-and-
material basis. In general, industrial firms are not externally
funded and thus their products must finance themselves!

The technological challenge is well accepted within industry, however, in the interest of the laboratories the competitive situation and the technological progress could well be improved if contracts are made to allow recuperation of actual costs. It is often argued that the technological spin-off will in future pay for today's investments in a new technology: This should not be overestimated since in most cases such a spin-off cannot be visualized and/or is too vague to be evaluated.

A main problem for industry is the lack of continuity of the magnet business. Most jobs require expensive tools and machinery as well as a highly qualified staff. Both machines and personnel are limited resources and cannot be adjusted to a drastically varying business volume on a short term basis.

A further problem frequently met with development projects is the transfer of know-how already present in industry or developed by industry. Contracts should allow for protection of such know-how thus enabling a controlled and/or restricted transfer, if requested.

Mutual benefit could be obtained, by supplying to the manufacturer as much background information as possible on a certain project. With this, manufacturers are in the position to propose solutions which may better fit the needs of the laboratory techniques and production facilities. This implies that industries should be involved in a very early stage of a project rather than squeezing industries in a short and thus ineffective tender period. Too short tender periods bare the risk that not the best solution to a technological problem is offered and that the scope of the task is not fully covered.

Nevertheless, the incentive to do magnet jobs will always be the fascination of a new technology and the technical challenge associated with it.

SUPERCONDUCTORS FOR ACCELERATORS AND DETECTORS

M. Thöner

Vacuumschmelze GmbH

Grüner Weg 37 6450 Hanau, F.R. Germany

Keystoned Rutherford type cables with NbTi and Nb_3Sn bronze route strands for beam line magnets were produced by Vacuumschmelze GmbH (VAC) in large production quantities as well as Al stabilized conductors for detector solenoids. The results of the production of 120 km of the HERA quadrupole cable will be shown. For LHC different kind of keystoned cable were delivered both with NbTi for the 1,8 K operating temperature regime and Nb_3Sn bronze route for operating at 4,2 K. Four 1 m model magnets were successfully tested. Detector magnets usually are built with AL stabilized conductors to allow sufficient radiation transparency. The large scale production of superconductors made by coextrusion for the projects ALEPH, CLEO II and H1 will be summarized.

For the quadrupole magnets of HERA/DESY Vacuumschmelze delivered 120 km of 23 strands keystoned cable (Fig.1a). The basic strand consisting of 636 NbTi filaments at about 20 μm diameter in Cu matrix were produced in a quantity of around 2760 km (Fig.1b). The mean critical current density of the strands was about 2500 A/mm^2 at 5 T and 4,2 K with a standard deviation of about 4%. By sorting the wire for cable production the standard deviation of the cable was reduced to ±1 1% with a mean I_c value of 8037 A at the sensitivity of 10^{-14} Ωm at 5,5 T and 4,2 K (specification 6962 A ± 2%). The critical current statistics of the cable measured by Brookhaven National Laboratory (BNL) is shown in Fig.2, allowing high field quality in the magnets. Magnetization measurements of the cables performed at BNL in low fields corresponding to injection condition show a mean value of 2 M = 41,2 mT at an external field of 0,3 T with a standard deviation of 1% (Fig.3). During cable production the geometrical tolerances were kept low on an acceptable level. The measured values are shown in Tab. 1 in comparison with the specified dimensions. The results of the cable production and the excellent properties of the tested magnets underline the suitability of the cables even in large quantities for pretentious beam line magnets like dipoles and quadrupoles.

New Techniques for Future Accelerators III
Edited by G. Torelli, Plenum Press, New York, 1990

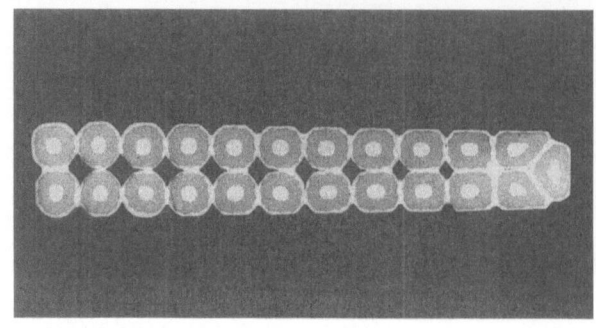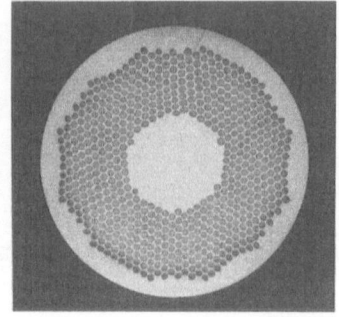

Fig.1a and 1b Cross-section of the 23 strands HERA quadrupole cable produced in total length of 120 km and of the strand with 636 NbTi filaments at about 20 μm diameter in Cu matrix

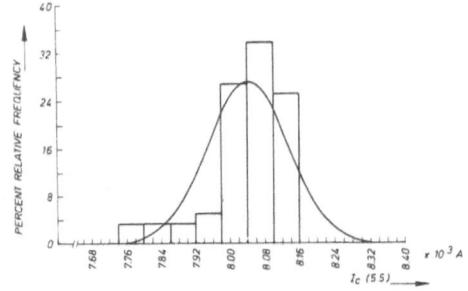

Fig.2. Critical current statistics of the HERA quadrupole cable with a standard deviation of ±1,1%

Fig.3. Statistics of the magnetization measurement of the HERA quadrupole cable at 0,3 T showing a standard deviation of ±1%

For the 1 m model magnets for LHC different kinds of key-stoned Rutherford type cables were delivered. First, for use of NbTi material at superfluid helium temperature, two tinned cables consisting of different strand material were used: cable I consisting of 30 strands with 1700 filaments and filament diameter around 10 μm. Cable II has 30 strands with 636 filaments and filament diameter around 20 μm. The strand diameter is in both cases 0,84 mm, the cable dimensions for cable I: 1,31/1,66 x 12,60 mm^2 and for cable II:1.31 /1,67 x 12,58 mm^2 (specification: 1,30 ± 0,02/1,65 ± 0,02 x 12,60 ± 0,02 mm^2).

Tab.1 Properties of the HERA quadrupole cable in comparison with the specification

Specification		Results during production
Stranddiameter	: (0,84 ± 0,01) mm coated with SnAg 5	uncoated: 0,83 mm coated with SnAg 5: 0,84 mm
Filamentdiameter	: ≤ 0,020 mm	approx. 0,018 mm
α	: $(1,8 ^{+0,2}_{-0,1})$	1,87 ± 0,03
Cable thickness (thin edge) a	: (1,33 ± 0,02) mm	(1,315 - 1,328) mm
Cable thickness (thick edge) A	: (1,67 ± 0,02) mm	(1,650 - 1,675) mm during cable production
Cable mid-thickness:	(1,50 ± 0,02) mm	(1,509 ± 0,005) mm resulting stack measurement
Cable width B	: (9,50 ± 0,03) mm	(9,470 - 9,504) mm during cable production
Twist pitch	: (70-90) ± 2 mm	88,3 mm
Number of strands	: 23	23
I_c at 4,6 K/5,5 T	: 6962 A ($\doteq$ 303 A/Strand)	I_c at 4,2 K/5,5 T (9877 ± 85) A (429 ± 18 A/Strand)
SC-Strands : 2.955 km $\doteq$ 68 billets $\doteq$ 262 single lengths		
SC-Cable : 125,2 km $\doteq$ 2.719 km SC-Strands (ca. 97 % SC-Strands worked down to Cable)		

Both magnets fulfill the design goals and reached considerable field level with excellent behaviour with respect to training, quench and reproduceability after heating up. Magnet model 8 TM1 reached 9,35 T at 10,53 kA, 8 TM2 9,45 T at 10,66 kA[1]. Second, for use of Nb3Sn material at 4,2 K, keystoned cables for a mirror dipole model magnet as well as the cables for a full dipole model magnet consisting of externally stabilized Nb3Sn strands made by the bronze route process (Fig.4) were delivered. This material based on Nb filament in CuSn matrix and has to be heat-treated after coil winding to create the brittle superconducting phase Nb3Sn ("wind and react"). The main properties of the inner and outer cables for the full dipole are given in Tab.2.

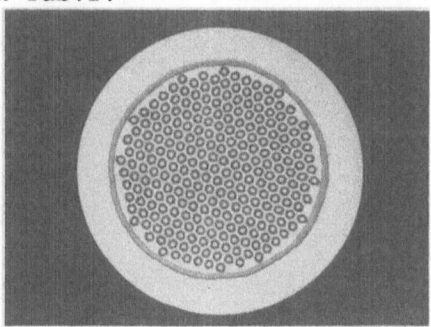

Fig.4. Cross-section of a typical externally stabilized Nb3Sn bronze route conductor

Tab.2 Specification of the Nb₃Sn cable for the 1 m dipole

	inner cable	outer cable
STRAND		
diameter	1,38 mm	0,92 mm
filament number	50 000	20 000
filament diameter	≈ 2,5 μm	≈ 2,5 μm
Cu stabilizer area	29 %	25 %
I_c (10 T, 4,2 K, 0,1 μV/cm)	849 A	395 A
CABLE		
strand number	24	36
width specified	16,81 ± 0,03 mm	16,81 ± 0,03 mm
delivered	16,813/16,819 mm	16,819/16,813 mm
thickness A specified	2,69 ± 0,02 mm	1,79 ± 0,02 mm
delivered	2,698/2,699 mm	1,781/1,774 mm
thickness a specified	2,19 ± 0,02 mm	1,47 ± 0,02 mm
delivered	2,201/2,204 mm	1,481/1,478 mm

Both, the mirror and full dipole magnet, were successfully fabricated, tested beyond the 10 T field threshold. The main dipole reached 10,05 T (central field 9,5 T), the mirror dipole 10,2 T at the conductor, both at 4,3 K [2]. These experiments have impressionably shown that the production of Nb₃Sn cable and manufacturing of magnets by the wind and react method is realistic alternative for beam line magnets in future accelerators.

Al stabilized superconducting flat cables for the solenoid of detectors were produced by coextrusion of high purity Al and cable by a continuous coextrusion process without length limitation by the process itself (Fig.5). The used modified cable sheathing press allows recharging of Al billets during coextrusion. The technique was developed to a standard fabrication process. The produced conductor length for ALEPH, CLEO

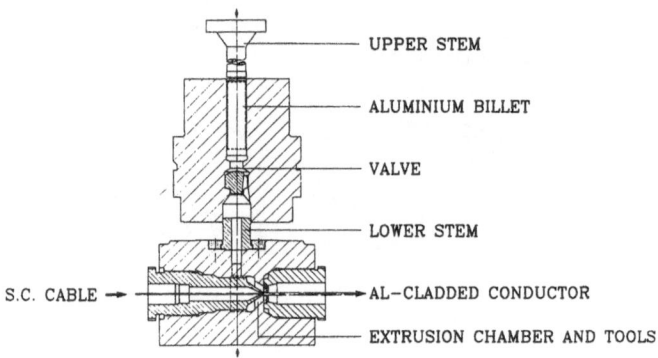

Fig. 5. Modified cable sheathing press for coextrusion of Al stabilized conductors

and Hl were about 70 km in unit lengths of 1400 m to 2200 m depending on the project requirements. The quality of the extruded conductors was measured by determination of the RRR value after extrusion, the shear strength and the contact resistance between cable and stabilizing Al. The shear strength, defined as the shear force devided by the surface area projection, was determined with specially prepared samples as well as the transfer length by determination of the voltage drop at samples with partially removed superconductors. Typical test samples are shown in Fig.6. All measured properties of the Al stabilized conductors were well above the specified values, a summarized overview of the three projects is given in Tab.3.

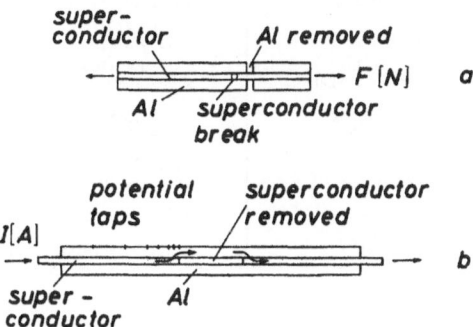

Fig 6a and 6b
Sketch of the experimental set-up for the measuring of the shear strength (6a) and the determination of the contact resistance (6b)

The coextrusion process was modified also for brittle reacted Nb3Sn-conductors. Many variants also with additionally coextruded reinforcement by high strengthening material were successfully produced in many hundred meters[3]. A typical reacted Nb3Sn cable coextruded with small Al cross-section and reinforcement profiles is shown in Fig.7 underlining the high potential of this process for future application.

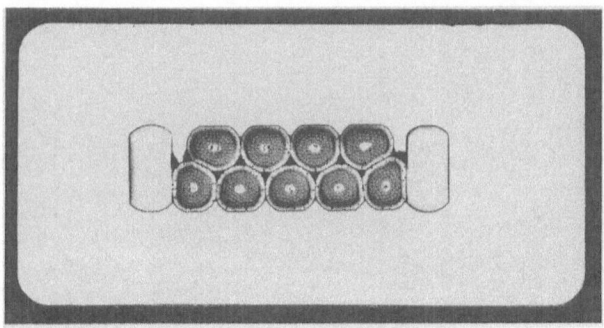

Fig. 7. Al stabilized Nb3Sn cable (3,29 x 1,18 mm^2) produced by coextrusion of high purity Al and profiles of high strengthening material (conductor dimension 8 x 4 mm^2)

Tab.3   Specification and test results of the aluminium stabilized conductors for the projects ALEPH, CLEO II and H1 in comparison

PROJECT		ALEPH	CLEO II	H 1
conductor cross section				
strands	type	FS 54	S 1	FS 54/FS 150
	diameter	0,8 mm	0,7 mm	0,71 mm
	number	16	9/11	24
dimensions	flat cable	(6,56±0,02 x 1,37±0,02)mm^2	(3,28±0,02 x 1,20±0,02)mm^2 (3,98±0,02 x 1,20±0,02)mm^2	(8,66±0,03 x 1,24±0,02)mm^2
	conductor	(35,00±0,15 x 3,60±0,10)mm^2	(16,00±0,10 x 4,89±0,03)mm^2 (16,00±0,10 x 5,10±0,03)mm^2	(26,00±0,25 x 4,50±0,05)mm^2
unit length (number of lengths)		2200 m (13) 1700 m (2)	1750 m (8)	> 1400 m (18)
I_c	spec.	≥ 12000 A/2 T	≥ 6500 A/0,75 T; 1,5 T	≥ 11250 A/3 T
	meas.	≥ 13500 A/2 T	8200 A/0,75 T 7380 A/1,5 T	≥ 15000 A/3 T
RRR	spec.	> 700	500 - 1200	> 500
	meas.	1700 - 2000	1000	> 900
"shear strength"	spec.	> 10 MPa	> 10 MPa	> 10 MPa
	meas.	15 - 18 MPa	30 - 40 MPa	18 - 29 MPa
contact resistance	spec.	-	< 10-10 Ωm	< 10-10 Ωm
	meas.	0,5-2,4 x 10-11 Ωm	0,4-3,2 x 10-11 Ωm	0,8-2,0 x 10-11 Ωm
coil manufacturer		CEA - Saclay	Oxford Instruments Ltd.	Rutherford-Appleton Lab.
final installation		CERN (LEP)	Cornell University	DESY (HERA)

REFERENCES

1.   R. Perin "First results of the high-field magnet development for the Large Hadron Collider", presented at MT 11, Tsukuba, Aug. 28 - Sept. 1, 1989

2.   A. Asner et al. "First Nb3Sn, 1 m long superconducting dipole model magnets for LHC break the 10 Tesla field threshold", presented at MT 11, Tsukuba, Aug. 28 - Sept. 1, 1989

3.   M. Thöner et al. "Aluminium stabilized Nb3Sn superconductors" Adv. Cryog. Eng. Mat. 34, 507 (1988)

COLD TESTS OF INDUSTRIAL PRODUCTION OF ANSALDO HERA DIPOLES

A. Bonito Oliva, R. Penco and P. Valente

Ansaldo ABB Componenti
Genova, Italy

INTRODUCTION

ANSALDO ABB COMPONENTI is going to complete the production of 242 HERA dipoles ordered by the Istituto Nazionale di Fisica Nucleare (INFN). A part of these magnets[31], before being closed in the iron yoke, must be tested in a liquid helium bath at about 4.6 K in order to have a statistical check of the production. The guaranteed nominal current In is between 7200 and 7400 Amps, depending on the short sample critical current Ishs (measured at 4.6 K and 5.5 T) and is evaluated using the formula:

$$In = (Ishs - 8000) /3 + 7200 \text{ A} \qquad \text{if } 8000 < Ishs \leq 8600 \quad \text{(A)}$$
$$In = 7400 \text{ A} \qquad\qquad\qquad\qquad \text{if } Ishs > 8600 \quad \text{(A)}$$

The production is divided in a pre-series of 20 and a series of 222 magnets. Ten preseries magnets and one magnet every ten of the series had to be tested. Up to now 26 dipoles have been tested. In this paper we describe the training observed on the dipoles to reach the guaranteed value of the current. Besides we present some characteristics of the coil behaviour during the quenches carried out.

TRAINING RESULTS

In Fig.1 the number of quenches occurred for each dipole before reaching the guaranteed current is shown. The distribution is rather spread; however it is possible to see a slight improvement in the production quality in the time. In order to analyze the quench current values obtained, it is interesting to make a comparison with the critical current. The critical current for the HERA cable, for a given temperature T and a given magnetic field B, is expressed by:

$$Ic (B,T) = Ishs \; 0.83^{(B-5.5)} (9.2 \; (1-B/14.5)^{.59} -T)/(9.2 \; (1-B/14.5)^{.59} -4.6)$$

where Ishs is the short sample critical current measured at B = 5.5T and T = 4.6K.

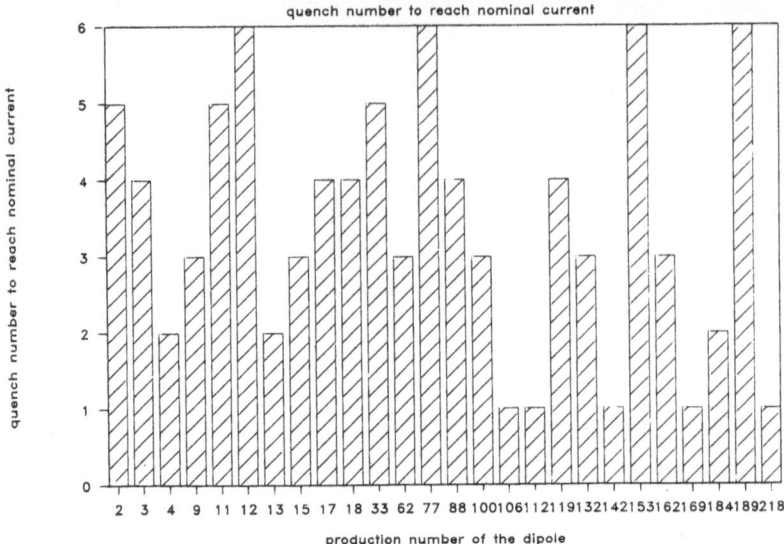

Fig.1 Quench number to reach the nominal current.

Using the formula: B = K*I for the maximum field on the conductor in the dipole, we obtain the critical current of the coil for every value of the current and temperature. Substituting the quench current and the experimental temperature, we can calculate the value of the fraction of the critical current for each quench. In order to carry out a statistical analysis of the training of the dipoles, we considered separately the first, the second and the following quench of each dipole. In Fig.2,3,4 the normalized distributions obtained for the dipoles are shown. On the abscisses the quench current-critical current ratio is reported; on the ordinates the normalized number of quenches occurred at a certain current is reported. This number is normalized with respect to the total number of quenches. As you can see, the first quench distribution is rather spread. The second is narrower and has a peak for I_q/I_c = 0.86. For the following quenches this effect is more marked and there is a peak for a bigger fraction of the critical current (I_q/I_c = 0.89). It is important to notice that we do not complete the training, since we stop the test when the coil reaches the guaranteed current.

QUENCH BEHAVIOUR

During the quenches we recorded the current, the current rate and the half coil voltages in order to calculate the coil resistance evolution.[1] We made tests discharging the coil on a 80 mOhm dump resistor (with and without quench heaters) or on the two power supply diodes. The most interesting phenomena observed is the increase of the quench speed during the discharging. The same phenomena has been also observed for discharging on the power supply diodes without quench heaters. The maximum field rate during a quench is typically 20 T/sec also for discharging without dump resistor: so the speed up could be due to the AC losses.

202

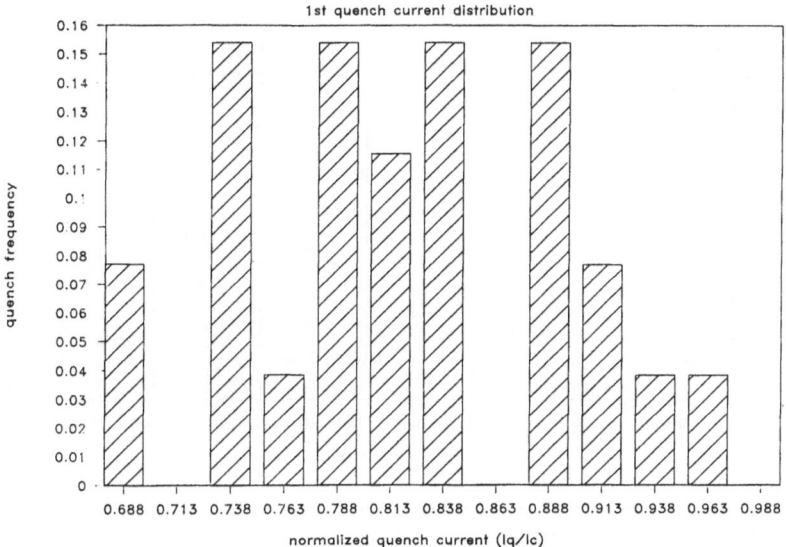

Fig.2 Quench current distribution for the first quench of each dipole.

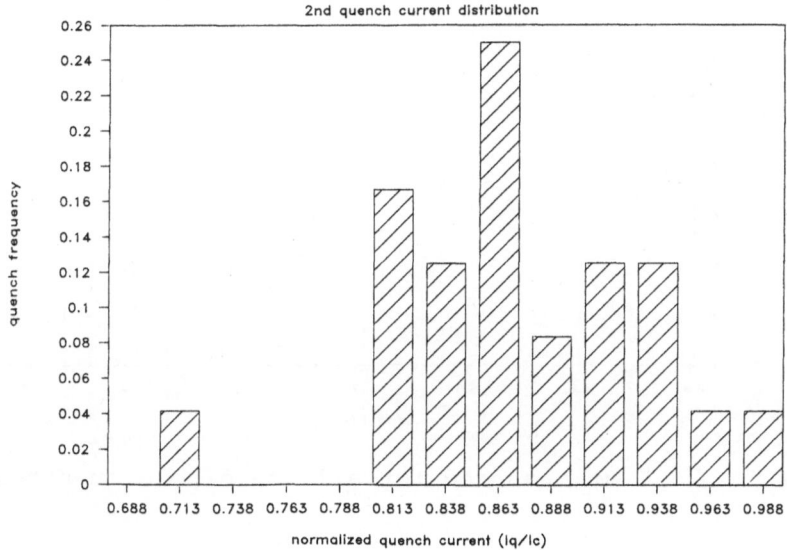

Fig.3 Quench current distribution for the second quench of each dipole.

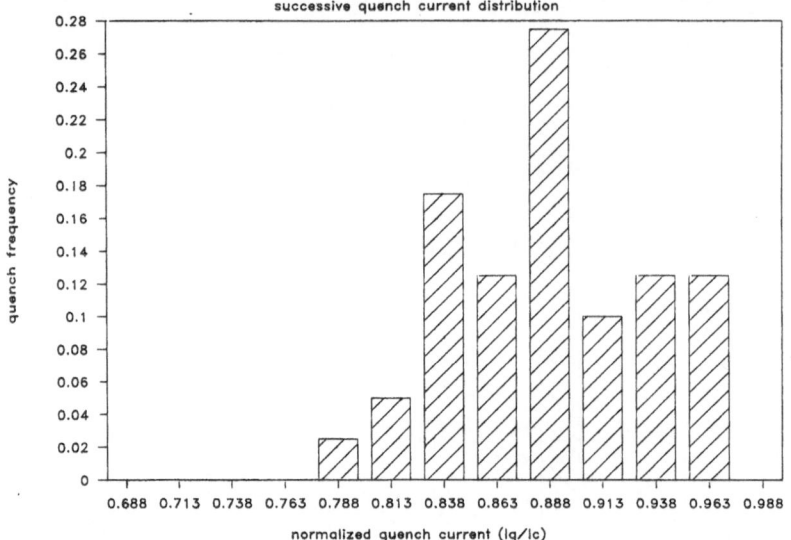

Fig.4 Quench current distribution for the successive quench
of the dipoles.

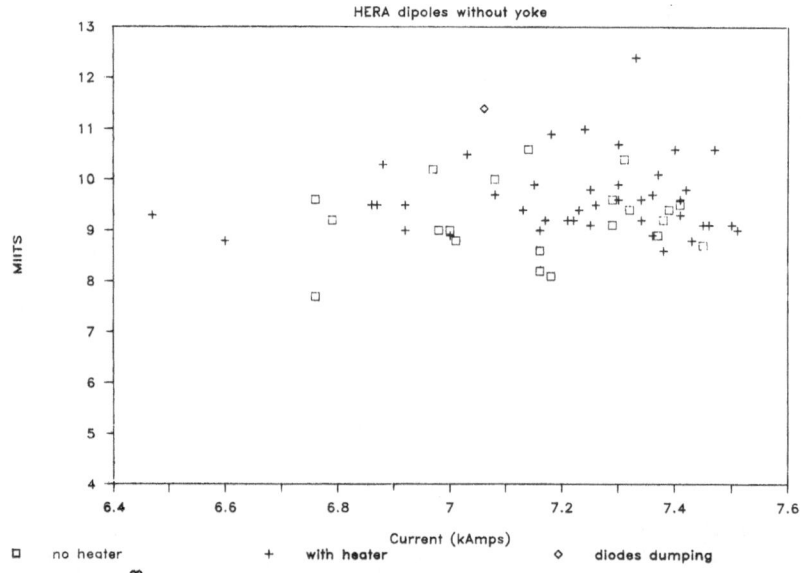

Fig.5 $\int_0^\infty I^2 dt$ measured for all the dipole quenches.

We calculated the value of $\int_0^\infty I^2\,dt$ for a quench by numerical integration. The values obtained for the quenches occurred for all the dipoles tested are shown in Fig.5. As you can see a negligible difference exists in the MIITS value for the cases with and without quench heaters. This is due to the very high quench propagation speed existing in the coil: when the quench heaters are powered, almost the whole coil is already in the normal state. It is interesting to analyze the point corresponding to the dumping on the power supply diodes without quench heater. There is a very little difference with the other points. This is due to the the fact that in a very short time the internal coil resistance becomes higher than the dump resistor value, so the two discharging times are very similar. This fact also means that the coil, for this current range, is self protected against hot spot problems.

REFERENCES

1. R.Musenich, A.Bonito Oliva, M.Losasso, G.Masullo, P.Valente, R.Penco, Quench behaviour of industrial HERA dipoles, presented at ICEC12, Southampton, UK, 1988

SUPERCONDUCTORS FROM FINLAND

Ian Campbell

Outokumpu Poricopper Superconductors

Pori, Finland

THE OUTOKUMPU GROUP

The Outokumpu Group is one of the world's leading base metal companies. The Group conducts mineral exploration, and mines and processes ore. It refines copper, zinc, nickel, cobolt products, precious metals, ferrochrome and produces stainless steel. Outokumpu further processes metals into finished products. It also develops and markets a wide range of equipment, electronics and know how.

In 1988 sales totalled some $2.5 billion. Exports and sales outside Finland account for 90% of the group's sales. Outokumpu ranks as one of Finland's major exporters.

The Outokumpu Group employs about 14,000 people, 5000 of which work outside Finland, in more than 20 countries worldwide. Over half of the production takes place in Finland. Since autumn 1988 Outokumpu's shares have been listed on the Helsinki Stock Exchange. The Finnish State holds 61% of the shares, private persons and institutions hold 39%.

OUTOKUMPU COPPER

Copper products is the Outokumpu Group's biggest business sector with primary customer groups in the construction, vehicle, electronics and electrical industries.

The business sector operates as a fully independent group under Outokumpu Copper Oy which includes Outokumpu Poricopper Oy in Pori, the Swedish Outokumpu Metallverken AB, The Nippert Company and Valleycast Inc., in the USA and 51% of Iberica del Cobre S.A. in Spain.

Outokumpu Copper operates through divisions according to product lines. Copper tubes, wire and drawn products, rolled products, radiator strip etc.

Sales in 1988 - $800 million, personnel 4669.

For years our list of customers has grown with our ability to meet critical specifications for a wide range of products. In resistance welding electrodes for the world's automobile manufacturing lines, magnet tubes for billionvolt particle accelerators and miles of hollow conductors for electrical motors and generators. Using electrolytic copper of LME (A) quality we cast billets for our special hot extrusion process which enables us to fabricate profiles, hollow conductors and rods for highly demanding applications such as x-ray tubes and semiconductor devices.

We are also the world's leading manufacturer of dispersion hardened copper spot welding electrodes for welding galvanised steel, now extensively used in the automobile industry, and to meet the needs of modern welding methods have developed special alloys of zirconium and chrome zirconium coppers. Over 50% of the cars coming off the USA assembly lines are now welded with our electrodes. Special manufacturing techniques provide smooth bore tubes for MIG/MAG welding - critical to the producers of long life, high quality welding tips.

Our upcasting technology is the basis for producing precisely dimensioned wires and strips, and is capable of producing small batches of specially formulated alloys. Upcaster units can service the needs of a remarkable range of applications from the mass production of zip fasteners to the very precise wires needed by the aircraft industry.

Machined copper products are fabricated mainly for the electroplating and metallurgical industries. Production includes copper anodes, continuous cast ingots, cooling elements and electrical bus bar systems. A chrome plating plant specializes in electrolytic coating and polishing of large components.

Our high quality NbTi superconducting wires and cables are used mainly are the production of magnets for particle accelerators, MHD energy conversion and NMR/MRI applications where specifications and quality are extremely demanding. Development of fine filament wires, filament size down to 2.5µm, is an important part of our R&D programme. The wires are excellent for Rutherford type cabling and have been used in particle accelerators.

As a significant producer of cryogenic oxygen free copper, and having within the company some of the most up-to-date process equipment, the superconducting department has control of its entire production. This gives flexibility, good quality control, shorter delivery times and competitive prices to its customers.

To meet the increasing demands of customers specific requirements we will continue to develop our manufacturing technology and processes, by ongoing R&D and so enable us to meet the almost endless applications for drawn and machined copper products.

Product types

Outokumpu superconductors are for specific purposes.

1.   Round and flat NbTi mono and multifilament wires with a Cu matrix for high homogenity persistent mode magnets and naturally for less demanding applications as well.

2.   Different types of cable made of the above mentioned wires, with or without copper and aluminium stabilizers.

HCOKOF

High Conductivity Outokumpu Oxygen Free Copper, the purest alloy of which is the cryogenic grade, is the best material for superconductors. Our skills in making this product - from purifying, melting and extrusion to the final wire drawing - have been acquired in more than five decades of work in this field. Outokumpu's own superconductor billet components assure the best semis that are totally free of impurities and particles.

Production facilities

The total production of superconductors - NbTi billet production excluded - is integrated in our copper division. The process begins with purification of the copper needed for the cryogenic grade copper billets, which are in turn the source of semis for the superconductor billets. Extrusion, drawing and heat treatment take place in a continuous production process. Integration of this kind permits flexible delivery times, and the all important quality assurance process can be carried out at every stage. Quality control is a totally scheduled system handled by an independent unit of the Quality Control Department of the Metal Laboratory.

Our metal-working facilities have been planned for the production of very big copper billets, and consequently long wire unit lengths are possible in superconductor production.

Testing

In addition to the considerable potential offered by Outokumpu's many laboratories and metallurgical Institutes, we have a well-equipped cryogenic laboratory for research and process control measurement. For testing we can use magnetic field of 10.5 T and direct current of 6000 A.

Service

We offer all the information and service of a large, multi-metal company. Intensive co-operation with customers will result in superconductors for the future.

MANUFACTURING OF SUPERCONDUCTING WIRES AND CABLES

G. Barani, S. Ceresara, G. Donati, R. Garrè

Europa Metalli-LMI Research Centre, Fornaci di

Barga, Lucca, Italy

ABSTRACT

The production of Europa Metalli-LMI in the field of traditional NbTi superconductors is described, with particular emphasis on the national and international projects the company has taken part in during the last few years.Basic research on A-15 compounds is also considered, with a survey of the various experimental methods developed for these materials.Finally, high temperature superconductors are examined, with reference to the studies carried out at EM-LMI Research Centre.

INTRODUCTION

Europa Metalli-LMI is the leading Italian company and one of the world leaders for the production of semifinished products in copper and copper alloys. In 1978 EM-LMI entered the field of superconductivity and since then it has taken part in both national and international projects, manufacturing a variety of NbTi superconductive cables. The development of advanced superconductors and research into high field materials are carried out by the EM-LMI Research Centre. A-15 compounds stabilized with copper are currently being investigated. In this respect, EM-LMI has recently been awarded a study contract by the NET Team, for a feasibility study and fabrication of a conductor prototype for the central solenoid of NET[1].

As for high Tc superconductors, basic research in this field is also in progress, aimed at the developing processing techniques suitable for industrial fabrication of superconducting ceramic wires or tapes.

2.NbTi SUPERCONDUCTORS

These materials are employed in applications in which magnetic fields of less than 9 T are needed. NbTi alloys, with a Ti wt % ranging from 45% to 52%, show a critical temperature of about 8 K and an upper critical field of about 14 T.

2.1 Wires

NbTi superconducting wires are manufactured starting from very pure Cu (RRR=250) and NbTi 46.5% alloy. These wires (called "Supermho"), are assembled in standard geometries including 60, 120, 540, 2550 filaments, both round (RW), and flat (FW). Critical current densities of 2700-3000 A/mm^2 at 5 T, 4.2 K have been achieved at EM-LMI on large scale production.

2.2 Cables

By suitably assembling Supermho wires, EM-LMI manufactures cables for very large transport currents in a variety of configuration (Supermho cables). Four typical examples of conductor geometries are given in the following.

- Fig.1 shows a cross section of a hollow superconducting cable used for the ENEA magnet operating in the SULTAN test facility at the SIN Centre near Zurich. The conductor, with external dimensions of 14.6 mm by 14.6 mm, internal dimensions of 8 mm by 8 mm, consists of 33 Supermho RW wires, 1.3 mm in diameter, wound around a copper tube and soldered to it. The diameter of the filaments is 50 μm, while the copper-to-superconductor ratio is about 20. The conductor carries a current of 12 KA at 5 T, 4.2 K.

- Fig.2 shows the geometry of the flat cable used for the conductor of the Italian INFN superconducting cyclotron. The conductor consists of a Supermho FW 540 wire, 3.6 mm by 1.8 mm in cross section, inserted and soldered into a copper slab, 13 mm by 3.5 mm in cross section. Filament diameter is 71 μm. The cable carries a current of 3 KA at 5 T, 4.2 K. EMLMI manufactured 26 Km of this cable.

- Fig.3 shows the Rutherford type cable employed for winding the superconducting dipoles of the proton ring of HERA at Desy (Hamburg). This conductor is composed of 24 "Supermho" RW 900 wires, 0.84 mm in diameter, while filament diameter is 16 μm. The cable carries a current of 8 KA at 5.5 T, 4.5 K. 500 Km of this cable have been manufactured by EM-LMI for winding 50 % of the dipoles.

- Fig.4 shows the Al-stabilized cable manufactured by EM-LMI used for winding the ZEUS magnet (HERA experiment). This cable consists of a Rutherford type cable, made of 10 Cu/NbTi composite wires, 1.04 mm in diameter (540 filaments), coextruded into a high purity Al jacket (RRR=1000), 15 mm by 4.3 mm in cross section. The (Cu+Al)/s.c. ratio is 14.6:1. The cable carries a current of 10 KA at 2.3 T, 4.2 K.

3. A 15 COMPOUNDS

Materials showing transport properties better than NbTi-based superconductors have to be used for applications in magnetic fields greater than 9 T. The EM-LMI Research Centre, in collaboration with ENEA has worked out original methods and procedures (2-4) for the fabrication of Cu stabilized A 15 superconductors. Concerning Nb$_3$Sn, the internal diffusion process

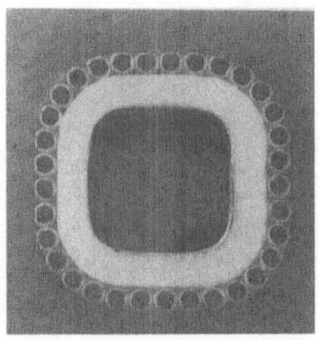

Fig.1 - SULTAN cable

Fig.2 - INFN cable

Fig.3 - HERA cable

A= 1,67 mm
C= 1,28 mm
B=D= 10,00mm

Fig.4 - ZEUS cable

is currently being studied. Fig. 5 shows the cross section of a
Nb_3Sn multifilamentary superconducting wire produced by EM-LMI.
Multifilamentary Nb_3Al superconducting wires have also been investigated for several years. Jelly roll, a particular fabrication process employed, results in filaments embedded in high
conductivity copper, a requisite of extreme importance for the
stability of superconductors. Fig.6 shows the cross section of
a multifilamentary wire before the heat treatment necessary for
the formation of the superconducting compounds.

4. HIGH TEMPERATURE SUPERCONDUCTORS

Since the discovery of high temperature superconductivity
by Bednorz and Mueller, Europa Metalli-LMI has been investigating ceramic superconductors. The research carried out at our
Research Centre can be summarized in the following four main
topics:

a) Development of new methods suitable for the preparation of high quality precursors. In this field we have
recently optimized a tecnique[5] allowing to obtain fine grained
stoichiometric powders (Fig.7).

b) Doping of YBCO to improve the physical properties of
ceramic superconductors.

c) Stabilization of ittrium based superconductors. By
working YBCO-based composite structures employing different
techniques, we are currently investigating the stabilization of
ceramic superconductive compounds with copper. In Fig.8 is
shown a semifinished sample, obtained by rolling, starting from
a composite structure in wich YBCO pellets were inserted into
holes obtained in an copper matrix.

d) Thick films prepared by electrophoresis. Considerable
activity is in progress in this field, aimed at the preparation
of very finely grained YBCO thick depositions with high critical current densities. In Fig.9 a prepared sample, deposited on
silver, is shown. Tc of about 90 K was measured.

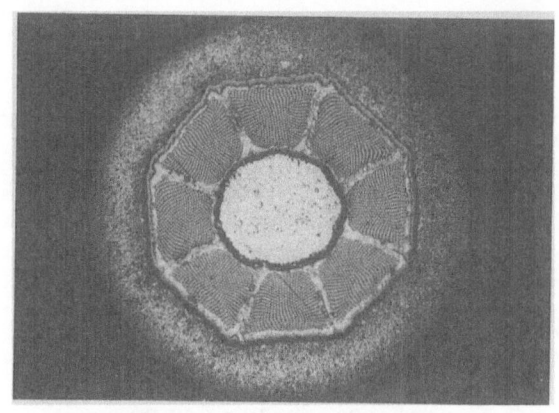

Fig.5 - Nb$_3$Sn wire

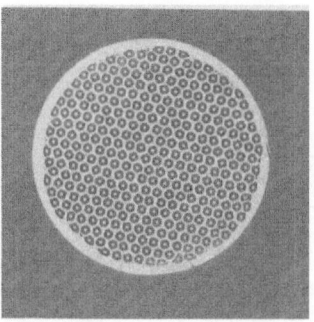

Fig.6 - Nb$_3$ Al

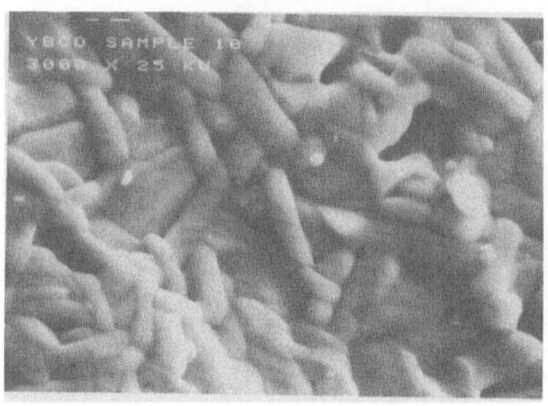

Fig.7 - SEM micrograph of
sintered YBCO

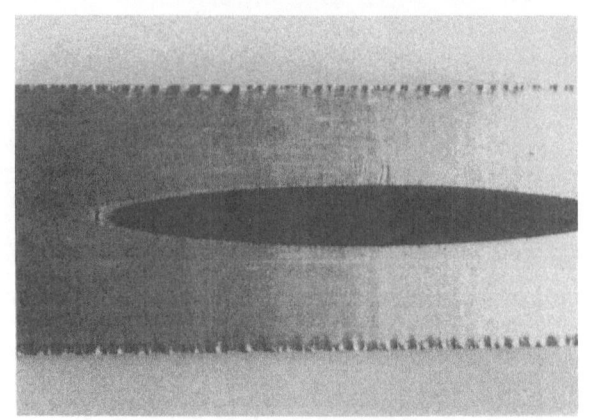

Fig.8 - Rolled YBCO

Fig.9 - YBCO thick film

REFERENCES

1) J.V. Minervini, N.Mitchell, R. Poehlchen, M. Ricci, E. Salpietro, A. Torrosian, P. Bruzzone, P. Marti, J. Rauch, P. Blasio, S. Ceresara, S. Rossi. Progress in the development of a 40 KA superconductor for the NET central solenoid. 11th International Magnet Technology Conference August 28-September 1, 1989, Tsukuba, Japan.

2) S. Ceresara, G. Sacerdoti G. and N. Sacchetti. Process for the Production of Superconductor Wires or Cables of Nb3Al and Superconductor Wires or Cables obtained thereby.USA Patent N. 559.628 (17th March 1975).

3) R. Bruzzese, S. Ceresara, G. Pasotti, M. Ricci, N.Sacchetti, M. Spadoni: Experimental Results on Nb3Al Multifilamentary Wires,VIth International Conference on Magnet Technology, Bratislava 28 Aug.- 2 Sept.1977. Proceedings (1978) pp. 1017-1020.

4) B. Annaratone, R. Bruzzese, S. Ceresara, V. Pericoli-Ridolfini, G. Pitto, N. Sacchetti: Effect of the thickness of Aluminium layer on the transport properties of Nb3Al superconducting wires.IEEE Trans. on Magnetics, MAG.17, n.1, pp.1000-1001 (Jan.1981).

5) N. Ammannati, G. Barani, R. Garrè, and S.Magnanelli. Preparation and Characterization of Y-Ba-Cu-Oxide Superconductor Obtained by Coprecipitation Method.Vuoto, vol.XIX,n.1, Genn.-Marz. 1989,pp. 24-27.

DEVELOPMENT OF A SUPERCONDUCTING SEXTUPOLE/DIPOLE CORRECTOR

MAGNET FOR LHC AT TESLA ENGINEERING, ENGLAND

S.A. Bates

TESLA

COMPANY PROFILE

Founded in 1973, Tesla Engineering has evolved into one of the most prominent designers and manufacturers of electomagnetic systems for high energy physics, medical imaging, space and defence.

Currently, 6000m^2 of modern, purpose built factory units exist, which are equipped with:

i) Automatic coil winding machines.
ii) Vacuum impregnation tanks.
iii) Machine shop which has 3 CNC milling machines.
iv) Automatic welding machine purpose built for yoke welding. The machine has four independently controlled welding guns, by cynergic welding plants.
v) Test facilities for measurement of coil electrical parameters, insulation quality etc.

The company is certified to defence standard AQAP 4 and adopts the same quality assurance procedures for all contracts.

LHC CORRECTOR MAGNET

The development of the superconducting sextupole-dipole corrector magnet was started by Tesla Engineering in January 1988. The first phase of the project was to construct a model made from ordinary copper conductor and subsequently to construct a prototype magnet from NbTi conductor.

MAGNET DETAILS

The sextupole is designed to correct for chromaticity as well as for sextupolar errors from other machine elements like the main dipoles and will operate between $-4000T/m^2$ and $+4000T/m^2$

The dipole will provide a means for orbit corrections and operates between -1.5T and +1.5T.

The LHC will require 1600 of these magnets and therefore, fabrication techniques have been chosen which are applicable to mass production. The development project is being carried out in the form of a cooperation agreement between CERN and Tesla Engineering, with the active participation of the Rutherford Appleton Laboratory.

GENERAL DESIGN

The magnet is constructed from the centre outwards in the following sequence :

i)   Sextupole coils
ii)   Dipole coils
iii)   Aluminium alloy shrink fitted collar
iv)   Iron yoke.

The overall yoke diameter is 170mm.

QUALITY REQUIREMENTS

Very high mechanical precision is required to reduce field errors to a minimum.

The radial position of the coil faces is required to be ±0.02mm.

In addition, the coil assembly has to act as a rigid, homogeneous mechanical body to stop quenching due to movement of the coils within the high magnetic fields present in the magnet.

The rigidity is obtained by final vacuum impregnation. A collapsable mandrel is used to impregnate the coils on in order that it can be removed after curing. The aluminium collar is then shrunk on to compress the assembly sufficiently to overcome shrinkage at the operating temperature. A shrink fit of 0.05mm is used.

SUMMARY

The development of the sextupole-dipole magnet is progressing well, given its inherent complications such as the extremely thin dipole conductor, 00.35mm, and the exceedingly tight mechanical tolerances.

Tesla Engineering have traditionally manufactured "conventional" electromagnets but are now diversifying into areas such as superconducting technology and permanent magnets particularly for undulators and wigglers.

The company has the ability to undertake many different types of magnet contracts, because most of the plant and equipment has been designed to be universal and easy to modify. This enables swift changeover which means delivery times can be extremely competitive.

Tesla is as keen to undertake small contracts as larger ones, since it recognises the need for strong links and cooperation with the scientific community. Technical expertise is gained in this way, both for industry and the research establishments.

TABLE 1

Design Parameters

Values of the prototype given in brackets.

	Sextupole Coils		Dipole Coils	
Field (gradient)	4000	T/m^2	1.5	T
Peak field in coil	4.2	T	3.5	T
Current	458	A	47	A
Number of turns/coil	104		1200	
Ampere turns/coil	47.6	kA	56.4	kA
Stored energy/magnet	6.2	kJ	5	kJ
Inductance	0.059	H	4.6	H
Inner radius	25	mm	36	mm
Outer radius	34.6	mm	40.2	mm
Length straight sectn	1	m	1	m
Operating temp.	1.8	K	1.8	K
	Sextupole Conductor		Dipole Conductor	
Overall section	0.7x1.2	mm^2	Diam 0.35	mm
Metal section	0.5x1.0 (0.58x1.08)	mm^2	Diam 0.3	mm
Filament size	25 (64)	μm	10 (21.5)	
Number of filaments	340 (54)		180 (36)	
Twist pitch	15 (25)	mm	5 (15)	mm
Copper/NbTi ratio	2/1 (1.6/1)		4/1 (4.4/1)	
Insulation enamel	Polyimide (PVA)		Polyimide (PVA)	
Breakdown voltage	2	kV	2	kV
Current density	915	A/mm^2	666	A/mm^2
Current dens. NbTi	2747 (2379)	A/mm^2	3332 (3596)	A/mm^2
% of short sample	59 (51)	%	65 (71)	%

REFERENCES

1.   R. Perin, A. Isjpeert (CERN) E. Baynham, P. Clee,
      R.Coombes (RAL), J. Wheatley, D. Willis (TESLA) - Report
      on the Development of a Superconducting Sextupole-Dipole
      Corrector Magnet.

PARTICIPANTS

Stephen BATES

TESLA Engineering LTD
Production/Technical
Water Lane Storrington
RH20 3EA W. SUSSEX, UK
Tel:*-903-743941
Fax:* -903-745548

Michele BARONE

INFN/LNF -Frascati
Tec. Division
Via E. Fermi, 9
00044 FRASCATI, Italy
Tel: 06-9403515

Louis BURNOD

CERN/SPS
CH-1211 GENEVE 23, Switzerland
Tel:*41-22-767-5146

Ian CAMPBELL

Outokumpo
Superconducting
P.O. Box 60
PORI, Finland
Tel: 010-358-39-826363
Fax: *10-358-39-827316

Aldo CATTONI

INF/LNF Frascati
Via Enrico Fermi, 40
00044 FRASCATI, Italy
Tel: 06-9403271

Cornells DAUM

Nikhef - H
CERN/EP Division
CH -1211 GENEVE 23, Switzerland
Tel: 0041-22-7672207
Fax: *41-22-787-9415

Oriano DORMICCHI

Ansaldo Abb Componenti
PMA/PRMA
Via N. Corenti, 8
16100 GENOVA, Italy
Tel:*39-10-6556455
Fax:*39-10-448937

Rens DUBBELDAM

HOLEC
Mechanical Development
Ringdijk 390
NL-2980-6BRIDDERKERN,The Netherland
Tel: *31-74-465573
Fax: *31-53-354003

Michele FERRO-LUZZI

CERN/EP Division
CH1217 GENEVE, Switzerland

Riccardo GARRE'

Europa Metalli LMI
S.P.A.
Centro Ricerche
Piazzale Orlando
55052 FORNACI DI BARGA (LU), Italy
Tel:*39-583-709137
Fax:*39-583-709567

Carl L. GOODZEIT

Brookhaven National Laboratory
BLDG 902B
NY-11973 UPTON, USA
Tel:*1-516-282-4711
Fax:*1-516-282-2170

Hermann ten KATE

Applied Superconductivity Center
University of Twenre
P.O.B. 217
7500 AE ENSCHEDE, The Netherlands
Tel:*31-53-893190/3847
Fax:*31-53-354003

K. JOHNSEN

c/o CERN
CH-1211 GENEVE 23, Switzerland

Gerard Jon KRAAIJ

ECN
Superconductivity Laboratory
Weste duirweg 3
NL-17552G PETTEN,The Netherlands
Tel:*31-2246-4085
Fax:*32-2246-4480

Rainer MEINKE

DESY/PMES
85 Notkerstrasse
D-2000 HAMBURG, FRG
Tel: *49-40-8998-3028
Fax:*49-40-8998-3094

Riccardo MUSENICH

INFN
Via Dodecaneso, 33
16146 GENOVA, Italy
Tel: *39-10-5993326
Fax:*39-10-308534
Fax::*609-921-3576

Thomas H. NICOL

Fermi National Accelerator
Technical Support/Engineering
P.O. Box 500
IL-60510 BATAVIA, USA
Tel: *1-312-840-3441
Fax: *1-312-840-3756

Romeo PERIN

CERN/SPS
CH-1211 GENEVE, Switzerland
Tel: *41-22-767-3285
Fax *41-22-782-2850

Paul REARDON

BNL/SAIC
Coll. Dek Center
NY-11923 UPTON, USA
Tel: *609-921-9030
Fax::*609-921-3576

Lorenzo RESEGOTTI

c/o CERN/LEP
CH-1211 GENEVE 23, Switzerland
Tel: *41-22-7672979

Giuseppe SCARFI

ANSALDO
Magnets
Via N.Lorenzi
16152 GENOVA, Italy
Tel:*39-6556601

Peter SCHMÜSER

DESY and Hamburg
Dept. of Physics
Notkestrasse 85
D-2000 HAMBURG 52, FRG
Tel: *49-40-8998-3884
Fax: *49-40-8998-3094

James STRAIT

Fermilab
Technical Support /Magnet R+D
P.O. Box 500
IL-60510 BATAVIA, USA
Tel: (312)-840-2240
Tel: *1-312-840-2240
Fax: *1-312-840-4756

Manfred THÖNER

Vacuumschmelze GmBH
R.a.D. of Superconductivity
Ehrichstrasse 5
D-6450 HANAU, FRG
Tel: *362-946
Fax: *362-980

Gabriele TORELLI

INFN Sezione di Pisa
Via Livornese, 582/a
56010 S. PIERO A GRADO, Italy
Tel: *39-50-960013
Fax *39-50-960676

Peter WANDERER

Brookhaven Lab.
ACC Dept. Bldg 902
NY-11973 UPTON, USA
Tel:*516-774-1112

Siegfried WENGER

ELIN CO/BA-PR
A-8160 WEIZ, Austria

Ferdinand WILLEKE

DESY/MPY
Notkerstr. 85
D-2000-HAMBURG 52, FRG
Tel: *49-40-8998-3197
Fax: *49-40-8998-3094

Stefan WIPF

DESY/F35H
Notkestrasse 85
D-2000 HAMBURG 52, FRG
Tel: *49-40-8998-3052

Jos VLOGAERT

CERN/SPS
CH-1211 GENEVE, Switzerland
Tel: *41-22-7675385
Fax: *41-22-782-2850

INDEX

Accuracy, 12
 angular _ , 14, 15
 geometrical _ , 13
 mechanical, 11, 12

Alignment, 50, 61, 63
 coil _ , 14
 quadrupole _ , 15

Aperture, 30, 70-72
 beam-pipe _ , 5, 16
 coil _ , 6
 dynamic _ , 3, 12, 25, 51,
 71, 72, 74-76, 78, 79, 82,
 84
 physical _ , 5, 79

Approximation, 50

Asymmetry, 12

Azimuthal field components,
 27, 35

Beam
  - beam collision, 128,
  130
 _ chamber, 115-117, 119,
  121, 125
 _ dynamics, 51, 80
 _ - gas interaction, 128,
  130, 144A
 _ injection, 44
 _ loss, 66, 127, 136, 143,
  146
 _ pipe, 16, 22, 28, 52,
  54, 59, 63, 66, 67
 _ stability, 72, 74

Betatron frequency, 71, 81
 _ oscillation, 69-74

BNL, 36, 59, 100, 147, 175,
 178, 195

CBA, 59, 183

CEA Saclay, 100, 104

Centering, 14, 16

CERN, 9, 66, 74, 84, 100,
 101, 104, 115, 123, 193,
 220

Chaotic, 71, 72, 74

Chromaticity, 25, 44, 52,
 53, 64-66, 69, 71, 78,
 79, 219
 _ correction, 3, 51, 63,
 66

Coils, 25, 27, 28, 30, 50,
 51, 56-58, 64, 66
 beam- pipe _ , 16
 _ alignment, 14
 _ aperture, 6
 _ clamping, 92
 _ cross-section, 12
 _ ends, 32
 _ heads, 50
 _ imperfection, 50
 _ package, 59
 _ winding, 13, 30
 correction _ , 16, 25, 28,
 57, 60, 66, 81
 multilayer _ , 54, 55, 59
 pick-up _ , 28
 radial _ dimension, 13
 quadrupole _ , 14-16, 35,
 59-61, 63
 sextupole _ , 16, 59-61,
 63, 82
 single layer _ , 54

Cold iron, 176

Cold mass, 15, 161, 163,
 166, 167, 172, 173, 176,
 180

Cold work, 190

Collars, 12, 14, 16, 18,
 19, 94-97, 149-153, 177,
 184, 220

Collider, 1, 49, 127, 130,
 148, 175, 176

Collimators, 127, 136, 137,
 143

227